Heidelberg
Science
Library

W0037639

J. Brachet · H. Alexandre

Introduction to
Molecular Embryology

Second, Totally Revised and
Enlarged Edition

With 77 Figures

Springer-Verlag
Berlin Heidelberg New York
London Paris Tokyo

Prof. Dr. JEAN BRACHET
Dr. HENRI ALEXANDRE

Université Libre de Bruxelles
Département de Biologie Moléculaire
Laboratoire de Cytologie et Embryologie Moléculaires
Rue de Chevaux, 67
B-1640 Rhode-St.-Genèse, Belgique (Belgium)

ISBN-13:978-3-540-16968-0   e-ISBN-13:978-3-642-82883-6
DOI: 10.1007/978-3-642-82883-6

Library of Congress Cataloging-in-Publication Data. Brachet, J. (Jean), 1909–   . Introduction to mo-
lecular embryology. (Heidelberg science library) Includes index. 1. Chemical embryology. I. Alexandre,
H. (Henri), 1943–   . II. Title. III. Series. [DNLM: 1. Biochemistry. 2. Embryology. 3. Molecular
Biology. QL 963.5 B796i] QL 963.B7313 1986 591.3′3 86-22111.

This work is subject to copyright. All rights are reserved, whether the whole or part of the material is
concerned, specifically those of translation, reprinting, re-use of illustrations, broadcasting, reproduction
by photocopying machine or similar means, and storage in data banks. Under § 54 of the German
Copyright Law, where copies are made for other than private use, a fee is payable to "Verwertungsge-
sellschaft Wort", Munich.

© Springer-Verlag Berlin Heidelberg 1986

The use of registered names, trademarks, etc. in this publication does not imply, even in the absence of a
specific statement, that such names are exempt from the relevant protective laws and regulations and
therefore free for general use.

2131/3130-543210

Preface to the Second Edition

Nearly 10 years have elapsed since I finished writing the first edition of *Introduction to Molecular Embryology*. During this period, molecular embryology has made great strides forward, but without undergoing a major revolution; therefore, the general philosophy and outline of the book have remained almost unchanged. However, all the chapters had to be almost completely rewritten in order to introduce new facts and to eliminate findings which have lost interest or have been disproved. There was a major gap in the first edition of this book: very little was said about mammalian eggs despite their obvious interest for mankind. Research on mammalian eggs and embryos is so active today that this important topic deserves a full chapter in a book concerned with molecular embryology. Therefore, I am very thankful to my colleague Dr. Henri Alexandre, who has written a chapter on mammalian embryology (Chap. 9) and has prepared all the illustrations for this book.

If, despite our joined efforts, the book is more ample and expensive than its predecessors, then this is the price that the reader will have to pay due to inflation: not only due to financial inflation that we all know only too well, but also to inflation in the publication of scientific works. Among them many deserve mention in the present book. Thus, the cost of scientific progress, for the authors, has been the necessity to extensively revise the first edition. But the real victim of the progress in molecular embryology is our devoted and efficient secretary, Ms. J. Baltus, who had to retype the entire book.

Chapter 2 has also been modified to a great extent: a description and schematic representation of a "typical" cell has been added since oocytes and eggs are cells, but specialized in a most important function, reproduction. The advent of new recombinant DNA technology now allows us to isolate almost any gene in amounts sufficient for detailed chemical analysis. This progress has led to entirely new ideas about the organization of DNA molecules. They are of the utmost importance for embryologists since the control of genetic activity is at the very heart of their problem: how does an egg give rise to an adult organism? As we shall see, a whole array of controls operate during *gene expression,* extending from the gene (a sequence of DNA bases located in the nucleus) to the cytoplasmic protein it encodes.

During the past 10 years, major progress in our understanding of oogenesis, maturation, fertilization and cleavage have been made; therefore, Chapters 4 to 6 had to be rewritten to a large extent. Progress has been slower in studies on the chemical embryology of invertebrates and vertebrates (except for mammals); but it is very encouraging that an ever greater number of gifted young scientists are tackling these problems with fresh ideas and improved techniques. A pleasant surprise is awaiting us in Chapter 10: it has been possible to obtain "transgenic" mice, thus fulfilling an old dream of both geneticists and embryologists. After injection of a pure gene into the nucleus of a fertilized mouse egg, mice which synthesize the protein corresponding to the foreign gene have been obtained; the gene and the capacity to synthesize the protein it encodes (human hemoglobin, for instance) can be transmitted to the offspring. Such new strains of mice, which bypass the slow and blind processes of Evolution, are a wonderful product of human skill and intelligence.

The molecular mechanisms of cell differentiation in embryos and in cultured embryonic cells have been extensively studied during the past years. Since the problems of cell differentiation, malignant transformation (cancer formation) and cell ageing are closely linked, the main results obtained in these very important fields during the last decade have been summarized in Chapter 11.

Another point, of more general interest, had not been stressed in the Preface of the first edition of this book: paradoxically, all living things display simultaneously both *unity* and *diversity*. Unity results from the fact that all of them obey the universal laws of molecular biology and biochemistry and that all living organisms are composed of cells which can multiply by division. Diversity is due to the fact that all living beings have different genes. Diversity in the living world is obvious when one compares animal or plant species using morphological criteria. It is also obvious in our own species since no man (except identical twins) has the same genetic make-up as another. The origin of genetic diversity lies in the earliest stages of embryonic development: exchanges of genetic material (genetic recombinations) take place at meiosis, thus during gametogenesis; acquisition of paternal and maternal hereditary characters results from the random collision of eggs and sperms at fertilization. The first stages of development (gametogenesis, fertilization, cleavage) display great similarities, despite quantitative differences, in the whole animal kingdom; as development proceeds, diversity becomes more and more apparent. Look, for instance, at the two cleaving eggs shown in Fig. 1: only an experienced embryologist can predict that they will give birth to organisms as different as a sea urchin and a mouse. After cleavage, complex cell movement, cell-to-cell interactions take place. Genes, which had remained silent during earlier stages of development, are now expressed in an orderly fashion, thus leading to different patterns in protein synthesis and to increasing morphological diversity. Development is thus the fulfillment of a *genetic program* inscribed in the nuclear genes (and to some extent in the cytoplasm) of the newly fertilized egg. Expression of these genes requires the co-operation of the egg cytoplasm, which possesses a high degree of specificity as a result of nucleo-cytoplasmic interactions during oogenesis. The genetic program inscribed in the fertilized egg chromosomes will continue to its end, unless accidents occur on the way: the ultimate stages of development are senescence and death. At the molecular, as well as the morphological, level, embryology is thus the science which studies the progressive and fascinating diversification of a small, almost homogeneous (at first sight) object, the egg; it allows us to see with our own eyes how apparent unity becomes diversity.

JEAN BRACHET

The second edition of this book, like the first one, is indebted to our students at the University of Brussels, to whom we have been teaching molecular embryology. We are very thankful to our colleagues H. Chantrenne, A. Ficq, R. Tencer and P. Van Gansen, who kindly read the manuscript and made many useful suggestions, and to Mrs. J. Baltus who typed it.

Brussels, August 1986   JEAN BRACHET
                                                         HENRI ALEXANDRE

Preface to the First Edition

The main questions that embryology has always tried to answer are the following: How can the fertilized egg, which has received a nucleus from the mother and another from the father, give rise to *all* the organs present in the adult? How is it possible, at a given time of development and in a special region of the embryo, that a limited number of cells can *differentiate* into muscles or red blood cells?

It is the purpose of *molecular embryology* to answer such questions in terms of the properties of macromolecules. Although we are still very far from obtaining a complete answer, spectacular progress in molecular biology has allowed us to state the same problems in simpler terms. For instance, the word "nucleus" can be replaced by "a set of genes" or still better, by "deoxyribonucleic acid (DNA)"; the fertilized egg has, in fact, received equal amounts of paternal and maternal DNA and this constituent is really the most important part of the nucleus, since it contains, in its own molecules, all the program for development into an adult. As for the second question, biochemistry tells us that muscle cells would fail to contract and would not display their characteristic structure under the microscope if they did not contain considerable amounts of specific proteins: the so-called "contractile" proteins, *actin* and *myosin*. A similar situation holds for red blood cells: they would not be red if they did not contain hemoglobin; without this red pigment, they would be unable to fulfill their mission, which is to transport oxygen to all other organs.

Our initial queries, when formulated in biochemical instead of morphological terms, become simpler. The fundamental problem of embryonic differentiation can now be stated as follows: Why is the DNA-directed ability to synthesize hemoglobin, actin, myosin, or any other specific protein expressed only at a given time and in a special region of the embryo?

Stated in this form, the question might, at first sight, seem more difficult than in its original form; in fact, it is an oversimplification of the real problem, but it is now easier to tackle experimentally. It is not enough to induce cells to produce hemoglobin (which can be done, as we shall see) to obtain their differentiation into red blood cells. If this book deals with *molecular* embryology, the reason is that we know enough about the mechanisms of protein synthesis and about their genetic control (see Chapter 2) to devise experiments that can throw light on the much more remote and complex problem of cell differentiation.

But, as we shall see, molecular embryology not only deals with embryonic differentiation, but also with all the basic phenomena of *cell biology,* such as cell growth, cell division, cell movements, nucleocytoplasmic interactions, and so on. In fact, a good deal of our knowledge about the chemical and molecular aspects of cell life comes from work done on very early stages of embryonic development. Sea urchin and amphibian eggs remain ideal objects for the study of cell division (mitosis) as well as cell differentiation.

This book will follow the very logical order of embryonic development, which is characterized by growing complexity. After introductory remarks about the history of molecular embryology and a short summary of our present knowledge of protein synthesis and its genetic control, we shall successively examine, from the molecular viewpoint, the main steps of development: the formation of eggs and sperms, fertilization, cleavage (which is a period of intensive cell divisions), early development of invertebrate and vertebrate eggs, nucleocytoplasmic interactions, and finally cell differentiation.

The general philosophy and outline remains the same as in *Biochemistry of Development* (Pergamon, 1960), a book I wrote, for a more specialized audience, more than 12 years ago. The present book owes much to my students at the University of Brussels, to whom I have been teaching, year after year, the progress of molecular embryology. It would not have been written without the kind insistence of Professor G. Czihak and Dr. K. Springer and without the efficient help of Professor P. Van Gansen (who took care of the illustrations), Dr. P. Malpoix (who revised the English), Mrs. J. Baltus (who typed the manuscript), and my colleagues H. Chantrenne, R. Thomas, A. Ficq, and P. Van Gansen (who read the manuscript and made many useful suggestions). My warmest thanks to all of them.

Brussels, July 1973   JEAN BRACHET

Contents

Chapter I   **From Descriptive to Molecular Embryology**   1

Chapter II   **How Genes Direct the Synthesis of Specific Proteins
in Living Cells**   7

Chapter III   **How Eggs and Embryos Are Made**   24

1   Theories of Embryonic Differentiation   24
1.1   Summary of Descriptive Observations   24
1.2   A Few Important Experimental Facts   30
1.3   Genetic Theories of Morphogenesis   34

Chapter IV   **Gametogenesis and Maturation:
The Formation of Eggs and Spermatozoa**   39

1   Oogenesis   39
1.1   Cytoplasm   39
1.2   The Nucleus (Germinal Vesicle: GV)   44
1.2.1   The Nuclear Membrane   45
1.2.2   Chromosomes   47
1.2.3   Nucleoli   51
1.2.4   Nuclear Sap (Nucleoplasm)   55
1.2.5   *Xenopus* Oocytes as Test-Tubes   56
2   Maturation of *Xenopus* Oocytes   58
3   Spermatogenesis   63

Chapter V   **Fertilization: How the Sleeping Egg Awakes**   66

1   General Outlook   66
2   The Fertilizin Problem   71
3   Physical and Chemical Changes at Fertilization   71
3.1   Early Changes   72
3.2   Later Biochemical Changes   72
3.2.1   Oxygen Consumption   72
3.2.2   Carbohydrate Metabolism   74

3.2.3   Protein Metabolism   74
3.2.4   Lipid Metabolism   74
4   Nucleic Acid and Protein Synthesis   75
4.1   DNA Synthesis   75
4.2   RNA Synthesis   75
4.3   Protein Synthesis   77
5   Parthenogenetic Activation   82
6   Molecular Embryology and Classical Theories of Fertilization . .   84

Chapter VI   Egg Cleavage: A Story of Cell Division   86

1   General Outlook   86
2   The Biochemistry of Cleavage   90
2.1   Energy Production   90
2.2   Chemical Nature of the Mitotic Apparatus   91
2.3   DNA Replication During Cleavage   95
2.4   RNA Synthesis During Cleavage   100
2.5   Protein Synthesis During Cleavage   103
2.6   Furrow Formation   104

Chapter VII   Molecular Embryology of Invertebrate Eggs   106

1   *Chaetopterus* Eggs   106
2   Mollusk Eggs   109
3   Tunicate (Ascidian) Eggs   110
4   Insect Eggs   113
5   Sea Urchins   115

Chapter VIII   Molecular Embryology of Amphibian Eggs   121

1   Respiration of the Amphibian Gastrula and Neurula   122
2   The Nature of the Inducing Substance   123
3   RNA Localization and Synthesis   125
4   The Links Between RNA Synthesis and Morphogenesis   130
5   Size and Mode of Action of the Inducing Agent   131
6   Molecular Basis of Cell Movements   132

Chapter IX   Molecular Embryology of Mammals   135

1   The Biology of Mammalian Sperm   135
1.1   Male Sex Determination in Mammals   135
1.2   Mammalian Spermatozoa from Testis to Fertilization   136
2   Molecular Embryology of Mammalian Eggs   139
2.1   Oogenesis   139
2.2   Maturation (Resumption of Meiosis)   140
2.3   Preimplantation Period   141

3   Experimental Embryology of the Preimplantation Stages   142
4   Early Postimplantation Stages   148
4.1   Trophectoderm Differentiation   148
4.2   ICM Development   150
4.3   X Chromosome Inactivation   152
5   Interspecific Hybrids and Chimaeras   153

**Chapter X   Biochemical Interactions Between the Nucleus and the
                Cytoplasm During Morphogenesis**   155

1   Biology and Biochemistry of the Alga *Acetabularia*   155
1.1   Biological Cycle. Regeneration   155
1.2   Morphology of the Cytoplasm and the Nucleus   158
1.3   Biochemical Studies   159
1.3.1   Morphogenetic Substances and mRNAs: Experiments with
        Inhibitors   159
1.3.2   Energy Production   160
1.3.3   Protein Synthesis   161
1.3.4   RNA Synthesis   162
1.3.5   DNA Synthesis. Relative Autonomy of the Chloroplasts   164
1.3.6   Complexity of Nucleocytoplasmic Interactions   165
2   Biochemistry of Anucleate Fragments of Eggs   166
2.1   Amphibian Eggs   166
2.2   Sea Urchin Eggs   167
2.2.1   Biological Observations   167
2.2.2   Biochemical Studies   168
3   The Importance of the Nucleus for Embryonic Development . .   171
3.1   General Background   171
3.2   Lethal Hybrids   172
3.2.1   Biological Observations   172
3.2.2   Biochemical Studies on Lethal Hybrids   173
3.2.2.1 Respiration   173
3.2.2.2 Nucleic Acid and Protein Synthesis   174
4   Biochemistry of Early Developmental Mutants   176
5   "Transgenic" Mice and Teratocarcinoma   178
6   Conclusions   180

Chapter XI   How Cells Differentiate   182

1   General Introduction   182
2   Specific Properties of Cell Membranes   185
3   Effects of Tissue Extracts on Cell Differentiation   187
4   Immunological Studies on Embryonic Differentiation   188
5   Enzyme Synthesis and Embryonic Differentiation   191
6   Differentiation of Cultured Embryonic Cells   193
6.1   A Brief Description of a Few Biological Systems   193

6.2   Experimental Analysis of Cell Differentiation in Culture   195
6.2.1   Cell Fusion   195
6.2.2   Effects of Bromodeoxyuridine (BrdUr) on Cell
        Differentiation   197
6.2.3   Effects of Phorbol Esters and Retinoids   197
6.2.4   Other Inducers and Inhibitors of Cell Differentiation   200
7   Embryonic Differentiation and Cancer   201
8   The Future of Molecular Embryology   203

References   206

Further Reading   209

Author and Subject Index   215

CHAPTER I

From Descriptive to Molecular Embryology

Embryology, the science of development, has like all other sciences its own embryology. How and when it was conceived remains unknown. Those very early days have been well described by the founder of chemical embryology, Joseph Needham, a gifted historian as well as a biochemist and embryologist. After man discovered artificial incubation of hen's eggs, which was used already in the oldest civilizations for very practical purposes, he must have had sufficient curiosity to break the shell immediately and look at the developing chick embryo. Without knowing it, he became a "descriptive embryologist" and studied morphogenesis (the appearance of new forms and functions). When he made the same kind of observations on a duck embryo, he became a "comparative embryologist". The first known descriptive and comparative embryologist was Aristoteles, who, more than 20 centuries ago, described the development of chick embryos and compared this development with that of embryos from fishes, reptiles, etc. Descriptive and comparative embryology remain at the root of modern embryology. These two branches of embryology, which grew so successfully in the eighteenth and nineteenth centuries, may now seem old; but they are by no means dead, especially in view of the continuous progress made in the optical means of observation. Thanks to the advent of electron microscopy, numerous and valuable papers have been published describing the ultrafine structure of normal sea urchin, frog or chicken embryos. This is the modern form of descriptive and comparative embryology.

However, after man had a good look at the developmental steps that lead to the formation of the adult, he became more curious than ever. How does this organism develop? In the eighteenth century, this question aroused great interest among natural philosophers: some held the view that the adult is already present, in miniature, in the egg (*preformation* theory). A minority, which was admonished by the religious authorities of those days, believed that embryogenesis is an *epigenetic* process; the embryo would form in a progressive way, as the result of complex dynamic changes. This battle was won by the epigeneticists and is now over; however, we shall, in this book, briefly discuss what preformation might mean in molecular terms.

After discussing at length preformation and epigenesis, embryologists were no longer satisfied with the mere observation of developing eggs and began experimenting on eggs and embryos: localized regions of eggs were killed by pricking, fertilized eggs were centrifuged, etc. *Experimental embryology* was born at

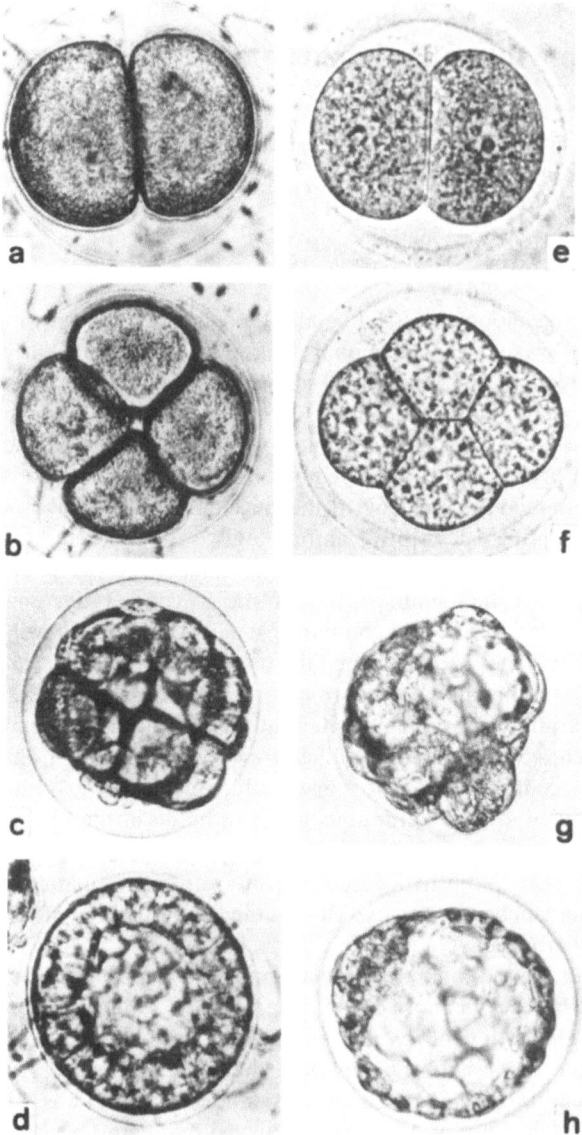

Fig. 1 a–h. Early development of sea urchin *Paracentrotus lividus* (**a–d**) and mouse (**e–h**) eggs. **a, e** 2-cell stage; **b, f** 4-cell stage; **c, g** 32-cell stage; **d** sea urchin blastula. **h** full-grown mouse blastocyst. (**a–d** Giudice 1973)

the beginning of the present century and culminated in the discovery of Spemann's "organizer": his experiments, which will be discussed in Chap. III, demonstrated that the nervous system of a frog does not result from the self-differentiation of a region of the embryo, but from interactions between this region and the underlying cells, i.e. these cells *induce* the differentiation of the undifferentiated ectodermal cells into nerve cells. In addition to induction, other important theoretical concepts, such as potentialities, determination, regulation, germinal localizations, morphogenetic fields and gradients progressively emerged.

Since *chemical embryology* is nothing more than the biochemical analysis of embryonic development, it naturally followed the same trends as embryology itself and its development has been necessarily linked to that of embryology. Chemical embryology began as a descriptive and comparative science and was later deeply influenced by the progress of experimental embryology. It is now being exposed to the impact of bacterial genetics and molecular biology. The use of new methods and concepts originating from molecular biology has transformed "chemical" into "molecular" embryology.

The first important landmark in the history of chemical embryology was Needham's *Chemical Embryology* (1931). Not only did Needham coin the name of the new science, but he assembled in three large volumes everything that was known at the time about the chemical composition of embryos at all stages of their development. This was the *opus magnum* of a great modern humanist and a bible for the small group of chemical embryologists who were just beginning research at that time. In hundreds of tables, data concerning the protein, glycogen, lipid and water content of all sorts of embryos were compiled from multitudinous papers. This compilation, which was so valuable at the time of its publication, has lost most of its usefulness now, since both problems and methods have undergone considerable changes.

Recent progress made in biochemistry and in physical chemistry had, of course, much influence on the orientation of the more modern work in the 1930s. While reinvestigating Warburg's former observations on the increase of oxygen consumption during fertilization in sea urchin eggs, Runnström introduced for the first time into these studies some very important biochemical factors (cytochrome oxidase, cytochromes, dehydrogenases and adenosine triphosphate). It was in the 1930s, too, that Needham, Chambers, Rapkine, Ephrussi, Wurmser and Reiss tried to measure the pH and the redox potential of these eggs. In view of the complexity of the ultrastructure of all cells, and eggs in particular (see Chap. II), we can no longer consider a cell as a water solution surrounded by a lipid membrane; it therefore seems foolish to try to determine the internal pH or redox potential of such a complex object as a whole, intact egg. Yet, as we shall see, changes in internal pH seem to play a very important role in the control of protein synthesis at fertilization. It was found by Chambers that the nucleus of an oocyte neither reduces nor oxidizes injected redox dyes; more recent studies have shown that the cell nucleus is indeed remarkably poor in oxidizing and reducing enzymes.

In 1932, a revolution occurred, namely the discovery by Bautzmann et al. that a killed organizer is still capable of inducing a nervous system after grafting it into an early frog or newt embryo. Biochemists discovered the existence of experimental embryology and the whole course of chemical embryology changed. This shift in interests is obvious when one compares Needham's *Chemical Embryology* (1931) with his *Biochemistry and Morphogenesis* (1942) or the senior author's *Embryologie chimique* (1944).

As soon as Holtfreter (1935) and Wehmeier (1934) found that the "inducing substance" (or "evocating substance" as the Cambridge group, following Waddington's suggestion, preferred to call it) was of very widespread distribution in Nature, the quest for its identification began. Since an alcohol-treated fragment of horse or ox liver, as well as a killed organizer, "induce" the formation of neural structures in the ectoblast, why not take the whole liver and fractionate it, following the biochemical methods which led to the isolation and purification of vitamins and hormones? This was done in several laboratories and the field was almost as "hot" as that of molecular genetics a few years ago. It soon appeared that every leading laboratory had its own "active substance", differing from that found in other laboratories.

But, suddenly, the setback came and the excitement ceased. In Cambridge, we (Waddington, Needham and Brachet) studied the respiratory metabolism of the organizer as compared with other parts of the gastrula. Since methylene blue was known to increase the respiratory rate of many cells, it was decided to implant pieces of agar containing methylene blue into gastrulae. Neural inductions were obtained, using an "inductor" of *nonbiological* origin. Still worse, neutral red, which has no effect on oxygen consumption, was also active. All the biological substances tested were more or less active; but it was impossible to decide which was the right one, since the ectoblast (like the unfertilized egg treated with parthenogenetic agents) could react in the same way (neural induction) to a large variety of chemical substances. This led Waddington et al. (1936) to suggest that the neuralizing factor was already present in the ectoblast, but in a bound form. A y agent which could "unmask" the inducing substance would induce the transformation of ectoderm into neural tissue. The "true" active substance would be liberated by the dying cells, and the hope of isolating and identifying it became remote. The final blow came when Holtfreter (1947) demonstrated that a short acid or alkaline shock, which is not sufficient to kill the cells, is sufficient to obtain a high percentage of nervous structures in ectoblast explants.

However, things are never as bad as they look. Discussions about the mechanisms of cephalic (neural) and caudal (mesodermal) inductions led to a resumption of work in a field which looked most unpromising since Holtfreter's observations. Were these differences due to quantitative or qualitative factors? Were there distinct neuralizing and mesodermalizing substances? It is to the credit of men like Toivonen, Yamada and Tiedemann that they have resumed the quest for specific cephalic- and caudal-inducing substances. Refined methods of protein chemistry were used (ultracentrifugation, electrophoresis, column chromatography) which led to the isolation from various tissues (liver, bone marrow, chick embryos) of purified active neural and mesodermal-inducing substances.

Isolation of specific, inducing proteins is part of molecular biology, which deals with the physical and chemical properties of biological macromolecules. But molecular biology has done much more than to provide new methods for studying the nature and the synthesis of the large molecules present in eggs and embryos. It has produced a revolution in our theoretical approaches, stemming from the tremendous development of molecular genetics. Molecular biologists, after having solved the riddles of heredity, are now entering the field of embryology and hope to solve the problems of cell differentiation. This massive injection of top biologists into this field will undoubtedly lead to spectacular progress.

This new trend can be exemplified in a field that has been the main battleground for molecular biologists: the control of specific protein synthesis by nucleic acids. In the "pre-Needhamian" period, around 1930, deoxyribonucleic acid (DNA) and ribonucleic acid (RNA) were unknown. Because of the Feulgen reaction, it was known, however, that "thymonucleic acid" (our DNA) is present in the nuclei of *all* cells, hence, no longer deserving its name of "animal" nucleic acid. Pentose nucleic acids (RNA) were supposed to be "plant" nucleic acids, although one of these acids had already been isolated from pancreas. No reliable method for estimating the two nucleic acids was available, since it was still believed that they could not co-exist in the same cell, one being specific for animal cells, the other for plant cells.

In the case of oocytes, the very presence of thymonucleic acid in the chromosomes was still hotly disputed; discussions between the holders of the "migration" theory and those of the "net synthesis" theory were intense. According to the former, the oocyte nucleus (germinal vesicle) contained a reserve of nucleic acid which moved into the cytoplasm at maturation and migrated back into the nuclei during cleavage. Their opponents, of course, denied the existence of such a reserve and claimed that thymonucleic acid was formed de novo in the nuclei after fertilization. The fact that the very unreliable biochemical methods then available for nucleic acid determination indicated that there was no change during development confused the situation even more. This constancy of the nucleic acid content during development was, of course, the main argument in favor of the migration theory; but the fact that the Feulgen reaction was negative in the oocyte and strongly positive in the nuclei of blastulae and gastrulae apparently demonstrated the correctness of the net synthesis hypothesis.

As usual, it was the development of a new and more specific technique (the diphenylamine method for DNA estimation of Dische) and of new hypotheses that clarified the situation. In 1933, the senior author demonstrated that unfertilized sea urchin eggs contain only traces of DNA as compared to the high DNA content of gastrulae. The reality of a net synthesis of DNA during development thus became an established fact. He also proved at the same time that eggs contain large amounts of RNA; therefore, the latter is not a plant nucleic acid at all, but is present in all cells.

A new question arose: where is the intracellular localization of RNA? In 1940, Caspersson's studies with the UV microscope and the senior author's utilization of a simple staining procedure (combined with the digestion of the tissue sections with ribonuclease in order to ensure specificity) brought the desired answer: RNA is an ubiquitous constituent of all cells. It is mainly localized in the nucleolus and the cytoplasm; and there is a close correlation between the RNA content of a given cell and its ability to synthesize proteins.

In vertebrate eggs (those of the amphibians, in particular), RNA is distributed along a primary, polarity (animal-vegetal) gradient already present in the oocyte and the unfertilized egg. At gastrulation, a new gradient, which is more dynamic and results from synthesis of fresh RNA, spreads from the dorsal to the ventral side. As a result of mutual interaction, dorsoventral and cephalocaudal RNA gradients become apparent in the late gastrula and the early neurula stages. These gradients are parallel or identical to the morphogenetic gradients of the experimental embryologists. Disorganization of the gradients by chemical or physical means and abnormalities in development go hand in hand.

The significance of the RNA gradients in amphibian eggs is becoming somewhat clearer. Autoradiography has now shown that RNA synthesis, which leads to the formation of the dorsoventral gradient at the gastrula stage, begins in the nucleus. On the other hand, electron microscopy has shown that the primary (polarity) gradient is essentially a ribosomal gradient. Combining these two findings we can arrive at the following explanation. The polarity gradient of inert ribosomes becomes activated toward gastrulation by messenger RNAs formed in the nuclei along the dorsoventral gradient. The result would be that protein synthesis will be more active in the anterior (cephalic) part of the embryo than in its posterior (caudal) part; protein synthesis would also decrease progressively from dorsal to ventral. Experiments with specific inhibitors of RNA synthesis (actinomycin D) and of protein synthesis (puromycin) confirm these deductions.

This is certainly progress; but to understand how production of specific messenger RNAs, controlled by localized gene activation leading to the synthesis of specific proteins, ultimately leads to the differentiation of nervous system, chorda or muscle cells is a task for the immediate future; it is a task for *molecular embryology*.

CHAPTER II

How Genes Direct the Synthesis of Specific Proteins in Living Cells

The control of specific protein synthesis is at the core of molecular biology; this problem can be studied in vitro in cell extracts containing enzymes which catalyze specific steps of this complex process. However, since we are dealing in this book with eggs and sperms, mention should first by made of *cell organization;* we shall then discuss specific protein synthesis in living cells. It should be emphasized at once that there is no such thing as a "typical" cell: as we shall see, both eggs and spermatozoa are highly specialized cells. During development, cells, which looked similar, undergo cell differentiation, a necessary step in the diversification of our tissues and organs (brain, muscles, skeleton, etc.). However, since all cells are built on the same morphological pattern, the simplified scheme shown in Fig. 2 can be used for didactic purposes.

All cells – except bacteria and mammalian red blood cells – possess a *nucleus* surrounded by a *cytoplasm;* they are limited by a *cell* (or *plasma*) *membrane* which regulates the exchanges of materials between two adjacent cells and with the outside medium. This membrane is composed of a double layer of lipids, in which "integral" proteins are inserted. Some of these proteins are enzymes which play an essential role in maintaining the cell's chemical composition. For instance, the plasma membrane contains a "sodium pump" which expulses excess sodium ions from the cell and allows a simultaneous intake of potassium ions. Outside the plasma membrane is the *extracellular matrix,* made of sugar-containing proteins (glycoproteins) and sulfated polysaccharides; its main function is to allow the cells to adhere to their neighbors or to an artificial substratum (glass, plastic, etc.). Just under the cell membrane is the *cell cortex:* its main constituent is a bundle of circular *microfilaments* made of *actin.* This protein can undergo linear polymerization or depolymerization by binding to other proteins; calcium ions are often required for binding between actin and the various actin-binding proteins, in particular to the contractile protein, myosin. The final result of this actin polymerization-depolymerization process is that cells can change their shape and undergo locomotion. In a tissue, cells are held together in close contact by various kinds of junctions, thus allowing the passage of small molecules and electrical current from cell to cell.

In the *cytoplasm* we find, as one of the main constituents, a *cytoskeleton* formed by a complex meshwork of fibrous molecules. There are three main cytoskeletal components: *microtubules* resulting from the linear polymerization of a protein called *tubulin;* actin *microfilaments* similar to those which are accumulat-

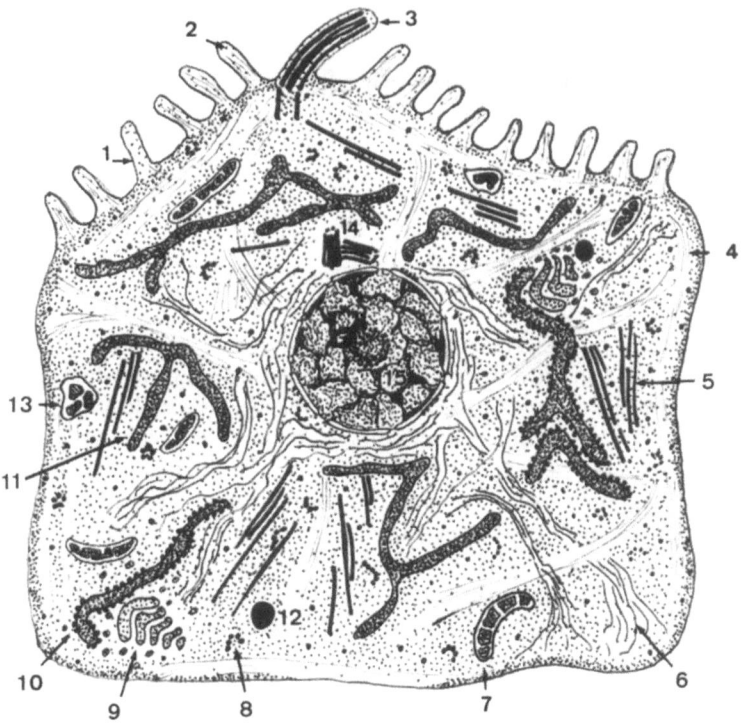

Fig. 2. A "typical" cell. *1* plasma membrane; *2* microvillus; *3* cilium; *4* microfilaments; *5* microtubules; *6* intermediate filaments; *7* mitochondrion; *8* polyribosome; *9* dictyosome (Golgi apparatus); *10* rough endoplasmic reticulum; *11* smooth endoplasmic reticulum; *12* primary lysosome; *13* secondary lysosome; *14* centrioles; *15* nucleus. (Drawn by Prof. P. Van Gansen)

ed in the cell cortex and which may display, if actin binds to myosin, contractility. *Actomyosin* is the major contractile protein of our muscles, but it is also present in almost all cells, including eggs. Finally, *intermediary filaments,* smaller in diameter than the microtubules and larger than the microfilaments, have a more diversified composition: they are made of fibrous proteins which differ in a tissue-specific manner. There are continuous interactions between the three kinds of cytoskeletal elements; the cytoskeleton is thus a dynamic, always changing, intracellular structure and lacks the rigidity of our own skeleton.

In the meshes of the cytoskeletal network, many organelles can be found. Of particular importance are the *mitochondria* and the *endoplasmic reticulum.* The former are surrounded by a double membrane and contain all the enzymes necessary for the oxidation of oxidizable substrates; their oxidation produces the energy required for the synthesis of macromolecules and for cell locomotion. However, cells cannot directly use the energy provided by cellular oxidations: this energy must be stored, in the form of energy-rich phosphate bonds, in a nucleotide called *ATP*. Its structure can be written as follows: adenine-ribose-phosphate~phosphate~phosphate, where the sign ~ represents an energy-rich bond. When the cell requires energy, the terminal phosphate bond is broken down by enzymes called *ATPases*. The *endoplasmic reticulum* is made of membranes (similar to the plasma membrane) associated to the *ribosomes,* which will be discussed later in this chapter; their function is protein synthesis.

The newly synthesized proteins can be either secreted out of the cell or retained inside the cell. Secretory proteins are produced by the endoplasmic reticulum, cellular proteins by free ribosomes (not attached to membranes). The role of the *Golgi bodies,* also found in all cells, is more complex; they contain enzymes which add sugar residues to the proteins synthesized by the endoplasmic reticulum. These sugars constitute "tags" for the various proteins which will be directed to their proper destination in the cell (as luggage is sent to its destination after being tagged in a railway station or an airport). Finally, we should mention the *lysosomes:* these acidic vacuoles contain hydrolytic enzymes which break down the large molecules which have penetrated into the cell by a process called *endocytosis.* For instance, if a foreign protein has been taken up into the cell by endocytosis, it will be broken down if it comes in contact with a lysosome; otherwise, it may remain intact for a long time. Endocytosis of proteins, which is a normal and important process, may be facilitated by the presence, on the cell membrane, of specific *receptors.* For instance, the pancreatic hormone insulin can penetrate into almost all kinds of cells because their membrane possesses insulin receptors which bind the hormone in a specific way; stimulation of cell division is often a result of the binding to receptors and subsequent endocytosis. Other cytoplasmic organelles, which are not found in all cells (centrioles, cilia, flagella, chloroplasts, etc.), will be discussed later.

The cell *nucleus* (Figs. 2 and 22) is surrounded by a *double nuclear membrane* filled with *pores;* more will be said about its structure and permeability in Chap. IV. The main constituent of the nucleus is *chromatin,* where *deoxyribonucleic acid* (DNA) is associated with proteins; morphologically, one may distinguish between a condensed *heterochromatin* and a loose *euchromatin.* The former is genetically inactive, i.e. it does not contain genes or, if it does, they are in an inactive form; its main role seems to be to keep the chromatin structure intact. In contrast, euchromatin contains active genes which direct, as we shall see, the synthesis of thousands of different proteins in the cell. In addition, most cells – but not all – have a variable number (in general, two) of *nucleoli.* These round bodies are made, as Caspersson and Brachet showed independently in the early 1940s, of *ribonucleic acid* (RNA) and proteins. Their function will be examined in Chap. IV. Recent work has further shown that the nucleus possesses, like the cytoplasm, a kind of skeleton. This is the *nuclear matrix,* a network of proteins which remains intact after the nucleic acids have been artificially removed from nuclei; long chromatin fibers are wrapped around this nuclear matrix.

After this very brief description of the cell morphology, it is time to discuss a few aspects of its biochemistry, in particular, how genes localized in chromatin direct the synthesis of proteins on cytoplasmic ribosomes.

It is impossible to do more than to summarize here, in a few pages, the essentials of the astounding discoveries made by molecular biologists during the past 40 years. The field is so large and is still moving so rapidly that only the basic facts required for the understanding of molecular embryology can be presented.

It has been known, for very many years, that *proteins* and *nucleic acids* are constituents of all cells, whether they belong to animals, plants or bacteria. Nucleic acids and proteins are also the main (and often the only) constituents of the viruses. These nucleoprotein particles can, if they are introduced into a living cell, replicate themselves repeatedly. When they do so, they retain all their genetic characteristics. If virus A_x is introduced into a suitable cell, many copies of the same A_x viral particles will be formed. This means that viruses, like cells and organisms, have *genetic continuity,* which is the main characteristic of life. The molecular basis of genetic continuity, as we shall now see, always lies in nucleic acids. For this reason, we cannot imagine any form of life, even very primitive, without nucleic acids and proteins. Their main functions are *replication* (one molecule of nucleic acid produces two identical molecules of the same nucleic acid) and *control of protein synthesis.*

The *proteins* are made of amino acids (R—CH—COOH), linked together

$$| \atop NH_2$$

by peptide bonds (—NH—CO—) in order to form *polypeptide chains.* Since there are 20 different amino acids and since their position can take any place in the polypeptide chains, an almost infinite number of proteins could theoretically exist.

The molecular weights of the proteins may vary from a few thousand to several hundred thousand (usually between 10,000 and 500,000) daltons. The largest proteins are often made by the association of a number (usually 2 or 4) of smaller

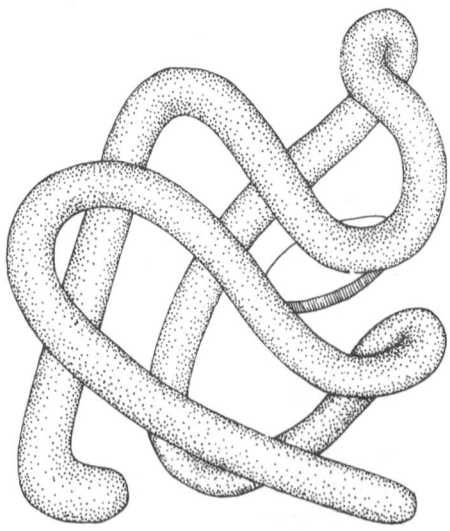

Fig. 3. Spatial (ternary) structure of the polypeptide chain of a hemoglobin molecule. (After M. Perutz)

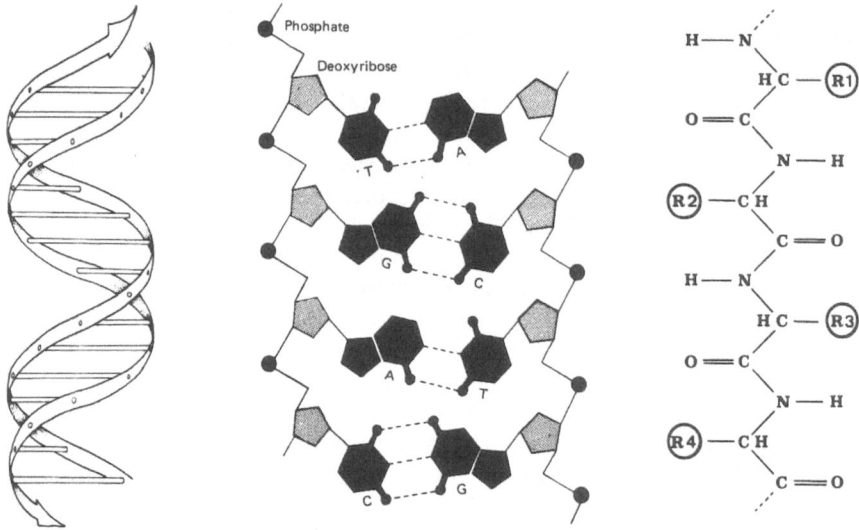

Fig. 4. *Left:* The DNA double helix, according to J. Watson, F. Crick and M. Wilkins. *Middle:* a segment of the DNA double helix; the bases adenine (*A*) and thymine (*T*) on one strand, and guanine (*G*) and cytosine (*C*) on the other, are linked by hydrogen bonds. *Right:* Segment of a polypeptide chain; *R1, R2, R3, R4:* side chains of the amino acids which form the polypeptide

subunits, which may be identical or different. The amino-acid sequence of the polypeptide chain defines the *primary structure* of the protein. But the polypeptide chain quickly folds and takes a tridimensional structure, which can be analyzed by X-ray diffraction crystallography. Present methods give us the possibility of establishing not only the amino-acid sequence of the peptide chains (the primary structure, Fig. 4), but also the precise spatial structure of such a complex protein as hemoglobin (Fig. 3). This molecule, which has a molecular weight of 68,000, contains four polypeptide chains which are identical in pairs (α_2, β_2). The hemoglobin molecule is abnormal and will not function properly if only one single amino acid of one of the polypeptide chains is replaced by another. Abnormal hemoglobins are far from rare in man, as a result of mutations which give rise to congenital diseases (a gene that has undergone mutation gives rise to an abnormal protein). It should be added that enzymes, which catalyze the biochemical reactions which occur in all cells, are always proteins. Recent research shows that they are often made of nonidentical subunits. One of them has the catalytic activity, while the other plays a regulatory role (it controls, in a positive or negative way, the catalytic activity, acting like the accelerator or the brake of a car).

Enzymes fulfill many functions in the cell: some of them, located in the mitochondria, are involved in energy production; others are essential for the synthesis and the breakdown of the macromolecules (nucleic acids, proteins, glycogen, etc.). Particularly important and interesting are the enzymes which exert a regulatory role in the cell metabolism: for instance, *adenylate cyclase* and *guanylate cyclase*, two membrane-bound enzymes, control the intracellular content of two cyclic nucleotides, cAMP and cGMP. In general, a high cAMP content inhibits cell division and favors cell differentiation; cGMP has the opposite effects. There is currently a great deal of interest in the *protein kinases*, which transfer phosphate from ATP to an acceptor protein; protein phosphorylation modifies the spatial structure and the electrical charge of the protein. Protein kinases, as a rule, require either cAMP, cGMP or calcium ions and phospholipids for activity.

Nucleic acids are, at first sight, much simpler molecules. Instead of 20 amino acids, they vary by the proportion of only four nitrogenous bases (two purines and two pyrimidines). The fundamental unit, in the nucleic acids, is the nucleotide, made of phosphoric acid, a pentose sugar, and a nitrogenous base. There are two main kinds of nucleic acids, which differ by the nature of their sugar moiety: the *deoxyribonucleic acids* (DNA) contain deoxyribose ($C_5H_{10}O_4$), while the *ribonucleic acids* (RNA) contain ribose ($C_5H_{10}O_5$). The two types of nucleic acids have three bases in common: the purines adenine (A) and guanine (G), the pyrimidine cytosine (C), while they differ in the fourth pyrimidine base; DNA contains thymine (T), RNA, uracil (U); but thymine is simply a methyluracil. Nucleic acids, like proteins, form long chains. They are, of course, not polypeptide, but *polynucleotide chains*. Figure 4 shows a comparison of a polypeptide and a polynucleotide chain.

For many years, noone thought that there could be a link between nucleic acids and proteins. In fact, as late as 1930, DNA was believed to be a small molecule, only present in the nuclei of animal cells, where it would act as a buffer. RNA was supposed to exist only in plant cells, where its possible biological role remained unknown.

A new era, that of molecular biology, started at the time of World War II when three important findings were made.

1. Cytochemical work by Feulgen, by Caspersson and by Brachet showed that DNA and RNA are constant constituents of all kinds of cells. DNA is primarily localized in the cell nucleus and in the chromosomes of the dividing cells. RNA is present in the nucleolus and in the cytoplasm of all cells. Furthermore, a close correlation was observed to exist between the RNA content of a cell and its ability to synthesize proteins. This correlation implied that RNA must play a major role in protein synthesis.
2. Avery discovered the phenomenon of *bacterial transformation*. He found that bacteria (pneumococci), unable to form a protective capsule, could build one if given DNA extracted from bacteria having such a capsule. Since formation or absence of a capsule is a hereditary characteristic, it followed that the genetic material was DNA. In other words, the genes, which are the chromosomal units of heridity, must be made of DNA.
3. The work of Ephrussi, Beadle and Tatum showed that individual genes control the synthesis of particular proteins. This is the "one gene–one enzyme" theory.

Since the gene is made of DNA and since an enzyme is a protein, it follows that the synthesis of an enzyme (that is, the assembly of a specific polypeptide chain) must be controlled by a definite sequence of nucleotides in the corresponding DNA macromolecule. Since, on the other hand, RNA is necessarily involved in protein synthesis, one cannot escape what Crick has called the "central dogma" of molecular biology: *DNA makes RNA, and RNA makes protein.*

How this happens is now well established because of many experiments largely carried out on bacteria and bacteriophages (viruses which infect bacteria). We now know that DNA molecules are made of a giant double helix (Watson and Crick) (Fig. 4), where the bases adenine and thymine (A:T) and guanine and cytosine (G:C) are linked together by hydrogen bonding. This structure, as we shall see later, provides an easy explanation for DNA *replication* prior to cell division; it also explains *gene mutation,* which can be due to substitution of one base by another or any kind of defect which will interfere with perfectly accurate base pairing.

The $A+T/G+C$ ratio differs from species to species; it can even vary markedly, in the same chromosome, between different stretches of the DNA molecule. Certain sequences are repeated in tandem many times and constitute *satellite DNAs;* other repeated sequences are interspersed, in many copies throughout the whole genome, between *unique* sequences. It is thought that these unique sequences are those which direct the synthesis of specific proteins and thus correspond to structural genes; the repetitive redundant sequences ("selfish DNA") probably play a regulatory or structural role (Doolittle and Sapienza 1980; Orgel and Crick 1980).

Until recently, it was believed that the huge DNA double helix is remarkably stable except under two circumstances: prior to cell division, it replicates (see Chap. VI) and after an injury (for instance a single-strand break), it is repaired. DNA replication and DNA repair are, in general, carried out by different enzymes. But the recent advent of great technical advances in molecular biology, which cannot be discussed here, has modified our views about the structure and stability of the DNA molecules. Briefly, the methods now available allow us to isolate fragments of the giant DNA molecules which contain individual genes (unique sequences) together with adjacent regions of the DNA double helix. Pure genes can now be obtained in sufficient amounts to analyze their structure, that is to establish the exact sequence of their bases. This work, which is in full swing today, has already led to a number of surprises. For instance, the discovery of „*mobile genetic elements*" in yeasts, maize, the insect *Drosophila* and probably many other organisms: these DNA sequences, which display similarities with the cancer viruses and with some of the interspersed repetitive sequences, can jump from one chromosome the another; their function, if any, is not yet clear. However, it is known that insertion of a mobile genetic element into a gene induces its mutation. The very existence of these elements shows that the classical view that the DNA content of a nucleus or a chromosome is absolutely constant can no longer be held. Sequencing DNA molecules has led to two other unexpected discoveries. For many genes, the region upstream the origin of the gene (that is, to the left of the gene) contains *regulatory sequences* which promote the transcription of the gene in the corresponding RNA sequences. However, in genes coding for small RNA molecules, the promoter of transcription is not located in regions flanking the gene, but in the middle of the gene itself (see Chap. IV for details). Another unexpected finding has been the discovery of *pseudogenes,* sequences which are very similar to those of a gene, but which cannot direct the synthesis of a protein because they are slightly abnormal. There is growing evidence that at least some of these pseudogenes originate from nuclear RNA molecules that have been copied into their DNA counterparts (reverse transcription); if so, genetic information does not always flow from DNA to RNA as required by the "central dogma" of molecular biology. For some pseudogenes, it is apparently RNA which has made DNA during Evolution.

There is increasing evidence for the view that Evolution results from deeper changes in the DNA molecules than the classical "point" mutations (changes at the level of a single base). Base sequencing shows that gene duplication, insertion or deletion of large DNA regions have been major factors in the production of mutations. Such events still occur today, as shown by the insertion of mobile genetic elements into many *Drosophila* chromosomal loci. In *Trypanosomes,* when the infected host raises antibodies against the surface glycoproteins of the parasite, a new surface coat protein entirely different from the previous one is synthesized. This happens because the parasite is capable of duplicating the gene which directs the synthesis of this new surface protein; this extra copy of the gene disappears when it is no longer required.

We know how the information contained in the "structural" genes, which are also called *cistrons* (that is, those unique sequences which direct the synthesis of a given protein or, more precisely, of a given polypeptide chain; the hemoglobin genes, for instance) is used for the synthesis of the corresponding specific protein. This information is *encoded* in the DNA molecule in the form of sequences of three nucleotides (*triplets or codons*). The information contained in the DNA molecule must first be *transcribed* into an almost identical RNA molecule; this is the so-called *messenger RNA* (mRNA), which is a faithful copy of one strand of the DNA gene. The copy is made in the cell nucleus by an enzyme called RNA polymerase II, which synthesize RNA only in the presence of DNA, which acts as a *template.* The mRNA is then released in the cytoplasm, which is the main site of protein synthesis; it contains the same information as DNA, in its codons. For instance, the AAA sequence (three adenines following each other) corresponds to the amino-acid lysine.

Thanks to the *genetic code,* which has been completely deciphered, it is now possible to deduce the amino-acid sequence of a protein from the base sequence of the corresponding gene. This type of work, as well as other experiments, has disclosed a surprising fact: the majority of the genes are "split", that is the coding sequences are interrupted by noncoding sequences; the first ones are called *exons,* the second *introns.* The gene coding for one of the collagens is fragmented

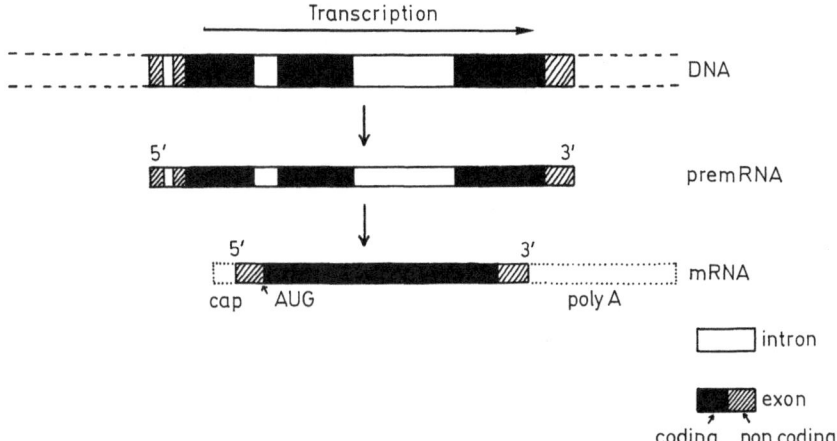

Fig. 5. The transcription of an eukaryotic gene. The gene (*DNA*) is interrupted by noncoding sequences (*introns*), which are transcribed and then removed during the maturation of the pre-mRNA into mRNA. In the mature mRNA, a *"cap"* (7 methyl-guanosine 5′ triphosphate, 7 mGppp) has been added at the 5′ end and a poly A sequence at the 3′ end; noncoding sequences are present at both ends of the molecule. *AUG* is the initiation codon for translation on the polyribosomes (see Fig. 6)

into as many as 50 introns; in contrast, a few genes have no introns at all. The significance of these mysterious introns is completely unknown; many think that they are just relics of the evolution of ancestral genes. The presence of introns obviously complicates mRNA synthesis (Fig. 5): the first product of RNA polymerase activity (the *primary transcript* or *pre-mRNA*) is a RNA copy of the whole gene, introns included. The transformation of this primary transcript into mature mRNA requires the removal of the RNA segments corresponding to the introns by a process called *splicing;* its detailed mechanism is not yet fully understood. It seems certain that the mRNAs cannot move from the nucleus to the cytoplasm unless the *maturation* of the pre-mRNA has been completed in the nucleus, i.e. when all the RNA copies of the introns have been removed.

Figure 6 summarizes, in a very simplified way, the mechanism of protein synthesis (translation of the genetic code). The specific mRNA molecules bind, in the cytoplasm, to already existing particles, the *ribosomes,* which result from the combination of four kinds of RNAs (the *ribosomal* RNAs or rRNAs) and a large number of ribosomal proteins. Ribosomal RNAs and proteins each form about 50% of the mass of a ribosome; it is customary to describe the ribosomes and the RNAs by their sedimentation constants (S) which can be measured in the ultracentrifuge.

Animal and plant cytoplasmic ribosomes have sedimentation constants of 80 S, while bacteria have smaller ribosomes (70 S). The cytoplasmic ribosomes, which are by far the most important site of protein synthesis, contain, in animal and plant cells, four different kinds of rRNAs, all rich in GC: 28 S rRNA, which has a molecular weight of about 1.3 million daltons, 18 S rRNA with a lower

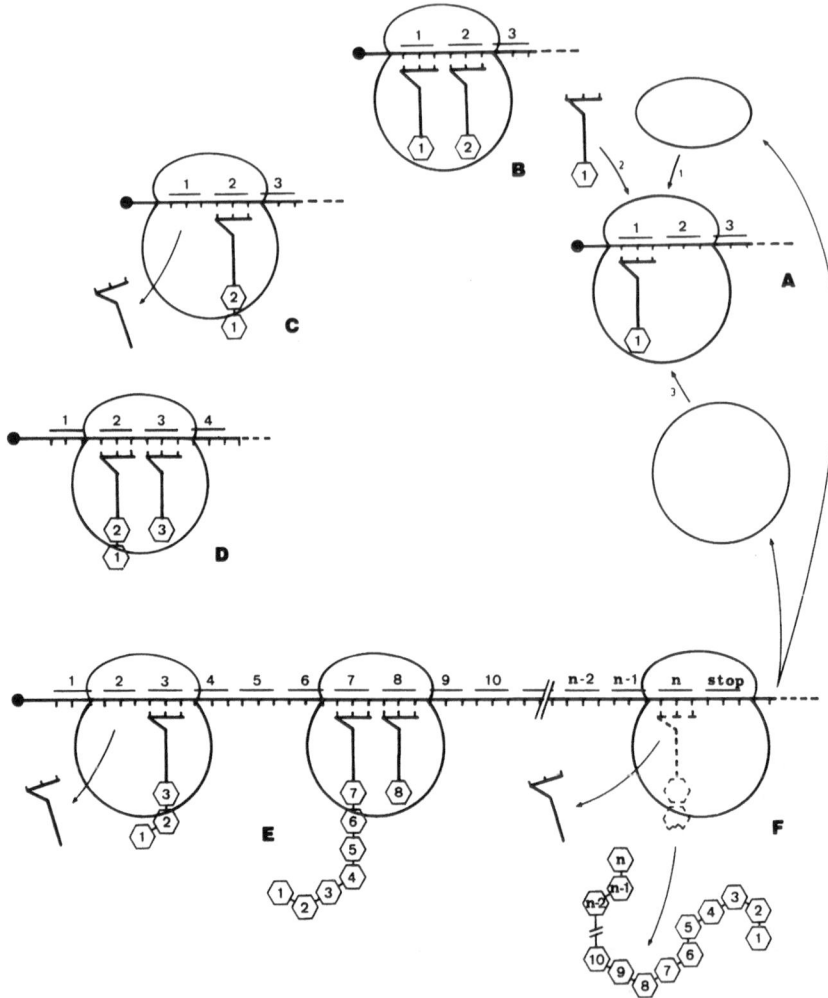

Fig. 6A–F. Protein synthesis. **A** Initiation: Binding of a small ribosomal subunit to a mRNA molecule (*1*). This is followed by the attachment to the AUG initiation codon in the mRNA of the corresponding aa-tRNA molecule (*2*) ([1]: formylmethionyl-tRNA). Finally, a large ribosomal subunit binds to the small subunit (*3*). **B** Attachment of a second aa-tRNA molecule [2] to the second codon of the mRNA molecule. **C** Formation of a peptide bond between aa₁ and aa₂ at the aa-tRNA-binding site. Ejection of the tRNA₁. **D** Movement of the mRNA in order to place the growing polypeptide at the "protein-binding site" and codon 3 at the "aa-tRNA-binding site". Binding of a third aa-tRNA molecule [3] to the third codon of the mRNA molecule. **E** Elongation of the polypeptide chain. **F** Ejection of both tRNAn and the synthesized protein

molecular weight of about 600,000 daltons, and the very small 5.8 S and 5 S rRNAs. Under appropriate circumstances, the 80 S ribosomes dissociate into two *subunits* with S values of 60 and 40, respectively. It is to the smaller 40 S subunit that mRNA binds, in such a way, that it links the ribosomes together to form larger aggregates called *polysomes*. There is very good reason to believe that association and dissociation of the subunits is an essential and necessary step in normal protein synthesis.

Why is the larger subunit made? It binds the representatives of the third large class of RNAs, the *transfer-RNAs* (tRNAs). These small RNA molecules (sedimentation constant of 4 S, molecular weight about 24,000) are now very well known since the complete nucleotide sequence of many of them has been firmly established. All tRNAs have the shape of a cloverleaf. At the center of the folded molecules is a specific sequence of three nucleotides, the *anticodon,* which binds with the corresponding triplet of the mRNA (the *codon*). If, for instance, the codon in the mRNA is AAA (which, as we have seen, codes for lysine), the anticodon will be UUU (since A and U bind together in the same way in RNA as A and T in DNA).

This particular tRNA will thus be a lysyl-tRNA, capable of accepting the amino-acid lysine at one end of its molecule; this end is made of the same nucleotide sequence (–CCA) in all tRNAs. There are enzymes in the cytoplasm that can remove or add this terminal –CCA sequence, with the result that the tRNA will be uncharged or charged with the corresponding amino acid. Because of the ribosomes and the tRNAs, the genetic message which has been transcribed from DNA to mRNA can be read and *translated* into protein.

It should be added that reality is still much more complex than this scheme. For instance, the genetic code is degenerate; this means that there can be several codons for the same amino acid. Consequently, several "iso-accepting" tRNAs can exist. They have different anticodons, but accept the same amino acid. The proportion and nature of the various iso-accepting tRNAs can change from cell to cell, and even in the same cell when, as the result of hormonal stimulation, for instance, it synthesizes a new protein. Another important factor is that, in order to synthesize a normal, "meaningful" protein, the synthesis should start with the right amino acids and stop when the last amino acid has been introduced in the growing polypeptide chain. Therefore, precise "signals" must exist that control the beginning and the end of the reading of the message. Nonsense codons, that is, nucleotide sequences which do not correspond to any amino acid, provide a means for stopping the growth of the polypeptide chain. The initiation of the latter is very complex indeed. There is not only an initiation codon (AUG) to which a specific tRNA (formyl methionine or methionine tRNA) binds by its anticodon, but at least three initiation factors of protein nature. Such protein factors (which are also needed for the dissociation of the ribosomes into subunits and for the elongation of the growing polypeptide chain) can be considered as specialized enzymes.

We have mentioned that all cells (except bacteria) contain *mitochondria,* which play a most important role in cell respiration and energy production. It is worth pointing out that the mitochondria behave, to a certain extent, like autonomous cell organelles. They contain their own DNA, which has been completely sequenced and is much smaller and entirely different from the main nuclear DNA. This mitochondrial DNA is replicated when the mitochondria divide and it can be transcribed by a mitochondrial RNA polymerase into all kinds of RNAs. The mitochondria contain their own ribosomes which are even smaller (55S) than those of the bacteria. These ribosomes are capable of independent protein synthesis, probably specialized in the building up of the proteins which form the mitochondrial membranes (structural proteins).

An important tool for the study of protein synthesis (which is very frequently used in molecular embryology) is the use of specific inhibitors of transcription and translation. *Actinomycin* is the best-known *inhibitor of transcription.* It binds to DNA and prevents the synthesis of all kinds of RNAs (mRNA, rRNA and tRNA), with a preference for rRNA (which represents 80% or more of the total RNA content of the cell). More recently, it has been shown that RNA synthesis can also be blocked with *α-amanitin.* This poison inhibits the activity of one of the RNA polymerases, RNA polymerase II, which is localized in the nuclear sap and plays an essential role in the synthesis of the mRNAs. The *inhibitors of translation* act at the level of the polysomes, thus in the cytoplasm. The most important are *puromycin,* which provokes the release of incomplete, inactive polypeptide chains and *cycloheximide* which blocks the movement of the ribosomes along the thread of mRNA and thus stops the "reading" of the genetic message.

Mitochondrial protein synthesis is hardly affected by these inhibitors, but is inhibited by the bacterial antibiotics *chloramphenicol* and *tetracycline.* Mitochondrial RNA polymerase, in contrast to the main nuclear enzyme, is unaffected by *α-amanitin,* but it is blocked by an antibacterial and antiviral agent, *rifampicine,* which has no effect on the nuclear RNA polymerases. As one can see, the use of proper inhibitors can allow a discrimination between the synthetic activity of the mitochondria and that of the rest of the cell.

Clearly, mitochondria have much in common with bacteria and it has often been suggested that they once originated from bacteria, and, in the course of evolution, adapted themselves to symbiotic life in animal and plant cells. It should be added that present- day mitochondria are almost entirely dependent on the rest of the cell for their survival and multiplication. They contain so little DNA, as compared to the nucleus, that their genetic information for protein synthesis is restricted to only a small number of proteins. Curiously, the genetic code used by the mitochondria is somewhat different from that of chromatin.

In fact, many of the respiratory enzymes present in the mitochondria are synthesized by the cytoplasmic polysomes under the control of nuclear genes, and move from the cytoplasm into the mitochondria.

More important than the mode of action of specific inhibitors of protein synthesis is the difficult problem of the *control mechanisms* of protein synthesis. We know, from the work of Jacob and Monod, how, in bacteria, gene activity is regulated by cytoplasmic repressors. A protein, the *repressor,* binds with extremely high affinity and specificity, to the DNA of an *operator* gene. When the latter is active (that is, its DNA is not blocked by the repressor), a whole battery of genes (the *operon*) becomes active, leading to the synthesis of the corresponding mRNAs and enzymes.

But it is certain that this scheme, which is valid for bacteria, is insufficient to explain the regulation of genetic activity in more complex cells or organisms. First of all, the nucleus of the bacteria is of the *"prokaryotic"* type; this means that it is extremely simplified. It is made of a circular thread of DNA, the bacterial chromosome, which is not surrounded by any membrane and is probably not associated with proteins. On the other hand, the nucleus of the *eukaryotes* is much more complex. It is surrounded by a *nuclear* membrane or envelope, it contains RNA-rich *nucleoli* and its DNA is associated with many different kinds of proteins to form the *chromatin;* the latter corresponds to the chromosomes of the dividing cells, but it is in a loose, uncondensed form.

There are two classes of chromosomal proteins, the *histones* and the *nonhistone* chromosomal proteins. One finds, in general, five main classes of histones, called H_1, H_2A, H_2B, H_3 and H_4; but certain cells, like the spermatozoa and the nucleated red blood cells contain other forms of specific histones or histone-like proteins. All the histones are small basic proteins with a high arginine or lysine content. All of them bind strongly to DNA and form particles with it called *nucleosomes* (Fig. 7). Histones H_2A, H_2B, H_3 and H_4 undergo polymerization, forming an octamer; a DNA fiber wraps around this histone *core*. Between two adjacent nucleosomes is a *linker* DNA, which may be associated with histone H_1. The length of the DNA molecule which surrounds the core is constant in all species and tissues. On the other hand, the length of the linker DNA may vary in a species- and a tissue-specific manner. The exact number of the nonhistone chromosomal proteins is not known, but it is certainly very high. Very important among them are the *RNA polymerases* I, II and III which are the agents of DNA transcription; they direct the synthesis of the large (28S and 18S) rRNAs, the mRNAs and the small RNAs (5S rRNA, tRNAs), respectively. These enzymes cannot work unless they bind to DNA; therefore, in order for transcription to start and to proceed, chromatin should be in a loose, "active" state accessible to the enzymes. We shall discuss later the mechanisms which control transcription; but it may now be mentioned that *active chromatin* differs from inactive chromatin in several respects. It is much more sensitive to deoxyribonuclease (DNase),

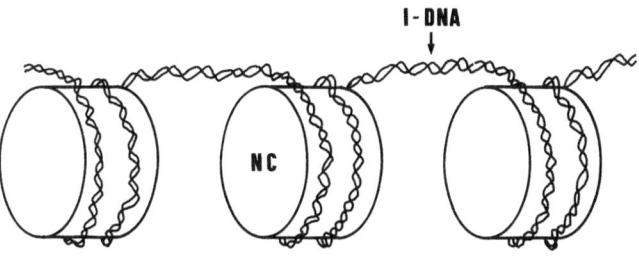

Fig. 7. Chromatin organization in nucleosomes. *NC* nucleosome core (histones); *l-DNA* linker DNA

the enzyme which breaks down DNA; it contains different amounts of basic non-histone proteins (called *HMG proteins*) which, like the histones and the RNA polymerases, bind strongly to DNA. In fact, many nuclear proteins are DNA-binding proteins; among them are enzymes called *topoisomerases* which modify the conformation of the DNA molecules. They contribute to the solution of a major problem for the cell nucleus, that of DNA compaction: DNA is a very long, but very thin molecule. How can it be packed into a nucleus? The first degree of compaction is achieved by the formation of nucleosomes, thus by the binding of DNA to core histones; a higher degree of compaction results from the coiling of the nucleosomes in order to form *supra-nucleosomal structures.*

Finally, it is very likely that the chromatin fibers form loops around the previously mentioned proteinaceous nuclear matrix. As one can see, the very important problem of the control of genetic activity is not yet solved; but we are beginning to see, despite all the complexities of chromatin structure, how to tackle and finally to solve this problem.

Protein synthesis is not only controlled at the transcriptional level. We shall find clear cases, in eggs, of *post-transcriptional* controls, which can occur at many levels: (1) in the nucleus itself, where unwanted mRNA molecules can be destroyed by hydrolytic enzymes (the ribonucleases); (2) at the nucleocytoplasmic border, where a selection between mRNAs which will function as messengers in the cytoplasm and others which will be retained and degraded in the nucleus probably occurs; (3) finally, at the level of the polysomes, in order to modify the translation of a given message. The conformation of the ribosomes, the binding of proteins to the mRNAs, the availability of tRNAs, of activating enzymes for the various amino acids, of initiation factors, etc., could all modify the amount and the nature of the proteins synthesized by a given cell. Since, as discussed in the Foreword, specific protein synthesis and cell differentiation are very closely related events, it is clear that any progress made in the field of gene regulation in eukaryotic cells is of primary importance for molecular embryology.

However, what we really want to know is still more than that: what causes the appearance of typical *morphological* structures, like the cross-striation of the muscle and heart cells, the production of cilia and flagella, which can beat and allow larvae and spermatozoa to swim, etc.? This is the basis of *cell differentiation,* which can easily be recognized under the microscope. It seems, at first sight, that we are dealing with such a formidable problem that improved knowledge of protein synthesis, of control mechanisms, of genetic activity, etc. will be of little use for its solution. And yet, there are good reasons for hoping that molecular biology and embryology will bring us a great deal closer to this still distant goal. Take the example of the cilia and flagella: they are made of a small number of proteins, which bear some similarities to the contractile proteins of the muscle. Like actomyosin, they contract in the presence of calcium ions and adenosine-triphosphoric acid (ATP), which is synthesized by the mitochondria during cellular oxidations. The proteins of the cilia and flagella (the major one is tubulin) exist in the form of rather simple elementary units. But they easily undergo linear polymerization; that is, they attach to each other in order to build

Fig. 8. Structure of a cilium (or flagellum) in an eukaryotic cell. *Left:* longitudinal section. *Right:* cross-sections of the cilium and its basal body (kinetosome). *1* ciliary membrane; *2* lateral arms; *3* the nine microtubule doublets; *4* central pair of microtubules; *5* central sheath; *6* radial spokes; *7* ciliary plate; *8* basal body (kinetosome)

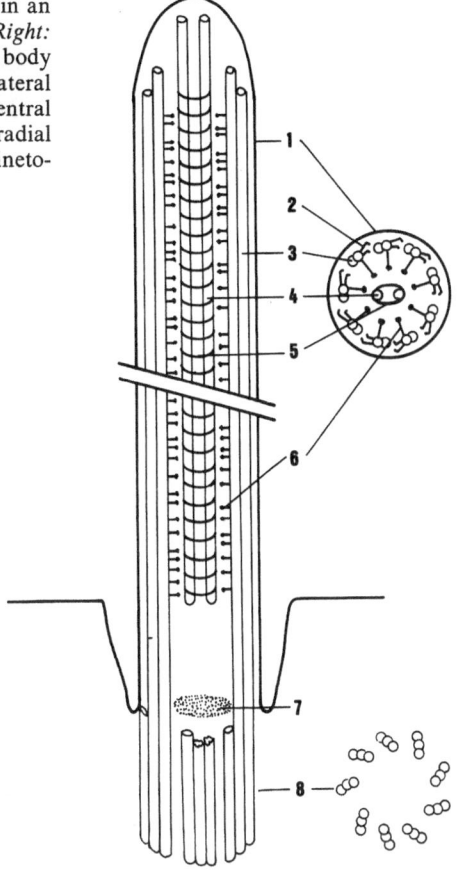

up elongate filaments similar to cilia and flagella (Fig. 8). These filaments can contract in the presence of ATP. In other words, it is possible to chemically dissect cilia into their elementary constituents and to reassociate the latter with the formation of structures which are morphologically, physiologically and biochemically very similar to cilia.

Such phenomena of *self-assembly* of protein units in order to form more complex structures are of course exceedingly important for molecular biologists. Still more striking are viruses, such as the DNA-containing bacteriophages and the RNA-containing tobacco mosaic viruses (Fig. 9). The infective genetic material of the virus is in both cases the nucleic acid. When a bacteriophage, which has the complex structure shown in Fig. 9 A, comes into contact with a bacterium of an appropriate species and strain, it injects its DNA into the bacterium. The protein coat does not participate in infection and plays the role of the syringe used for an injection (Hershey). The phage DNA immediately starts to synthesize the complementary mRNA and, after a series of very complex events, more phage DNA

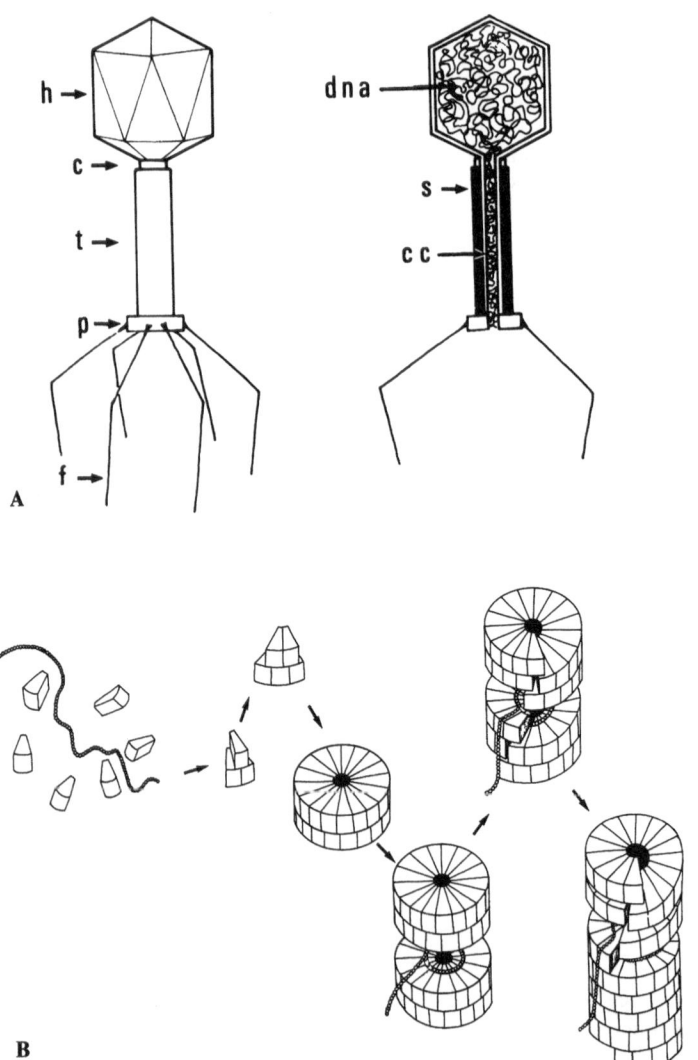

Fig. 9 A, B. Structure of two viruses. **A** Bacteriophage (T4-type). *Left:* schematic view of a virion; *h* head; *c* collar; *t* tail; *p* tail plate; *f* tail fibers. *Right:* ideal section through a virion: *cc* core with central channel; *s* sheath. *dna* DNA molecule coiled inside the head. **B** Self-assembly of proteins and RNA in tobacco mosaic virus. (From Klug 1972)

and the phage proteins are made. The head and tail proteins will assemble together, as in a jigsaw puzzle, in order to form the complex coat typical of the mature phage particles. The simpler tobacco mosaic virus is still better understood, since it is composed of viral RNA and a single protein only. If the viral RNA is introduced into the cells of a tobacco leaf, it will act as a messenger and induce the synthesis of the viral protein, using the ribosomes of the host cell for the formation of the polysomes. Once the viral protein has been formed, complete rods of virus particles will form, as a result of self-assembly of RNA and protein molecules (Fig. 9 B).

Studies on the self-assembly of viral nucleic acid and proteins represent a very useful model for the understanding of the much more complicated morphogenic events which occur during embryonic development. They also inspire the hope that it will be possible, in a not very distant future, to synthesize the RNA and the protein constitutive of the virus and to obtain in vitro, fully infectious tobacco mosaic viral particles.

CHAPTER III

How Eggs and Embryos Are Made

1 Theories of Embryonic Differentiation

1.1 Summary of Descriptive Observations

It would, of course, be impossible to reduce to a short summary all that has been learned about embryonic development in the multitude of animal species that have been studied from the descriptive and experimental viewpoints by scores of embryologists over a period of 200 years or more. However, a few of the most salient findings must be outlined as briefly as possible before we can proceed further to underline the molecular mechanisms at work during embryonic development.

Sexual reproduction is based on the formation of *gametes* (eggs and spermatozoa), which fuse together at the time of *fertilization*. All the cells of the adult originate from the fertilized egg; like the latter, they contain *n* chromosomes coming from the egg and *n* chromosomes brought by the spermatozoon. Thus, they contain two sets of homologous chromosomes and are said to be *diploid* (2*n*). The gametes, on the other hand, contain only *n* chromosomes and are said to be *haploid* (*n*). The reduction in chromosome number, which occurs at the end of gametogenesis, is brought about by a special kind of cell division, *meiosis*. The results of the two meiotic divisions, which follow each other at short or long intervals according to species, are shown in Fig. 10. A mother cell, spermatogonium or oogonium, grows in order to become a spermatocyte or an oocyte. In the male, the two meiotic divisions lead to the formation of four haploid spermatozoa; in the female, these divisions are unequal, and lead to the production of one haploid egg cell (ootide) only; the small polar bodies soon degenerate.

Description of *meiosis* (Fig. 11) can be found in all textbooks of cytology and embryology. It is characterized by a complicated prophase, which can be subdivided into a number of stages. The chromosomes, which have already replicated their DNA, become visible at the *leptotene* stage; the homologous chromosomes, of paternal and maternal origin, undergo pairing at the *zygotene* stage. They undergo strong condensation at the following *pachytene* stage. The homologous chromosomes tend to separate from each other at the *diplotene* stage; but they still remain attached together at certain points, called *chiasmata*. Finally, at *diakinesis,* the chromosomes again become thick and short. This long prophase is followed by the *first meiotic division,* which is different from the mitotic divisions.

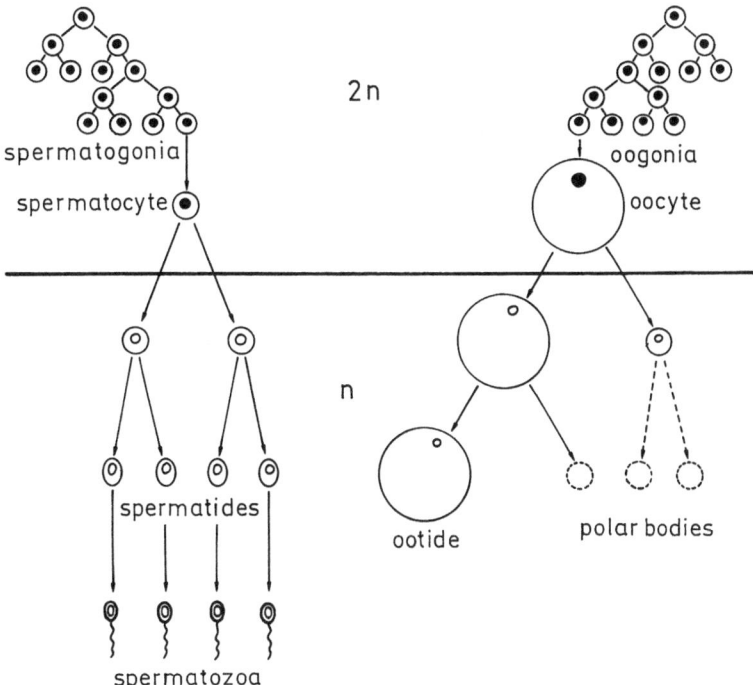

Fig. 10. Gametogenesis. *Left:* spermatogenesis. *Right:* oogenesis

The chromosomes do not divide into two daughter chromosomes, so that the two resulting cells are haploid (*n* chromosomes) and not, as usual, diploid (2*n* chromosomes). The main importance of meiotic prophase is the following; the precise pairing of the homologous chromosomes allows precise exchanges of genetic material between the homologous chromosomes (crossing-over), leading to genetic recombination. These exchanges require breaks in the DNA molecules, which are repaired by DNA repair enzymes.

Gametogenesis and *fertilization* will be discussed in more detail in the following chapters. It should be pointed out here that both the egg and the spermatozoa are haploid cells, but they are very different in size.

The egg is full of reserve materials (glycogen, fats, yolk proteins) which are responsible for its huge size. The tiny spermatozoa move actively because of their flagella. They are, as we shall see, machines adapted, like the bacteriophages, to the injection of the intact genetic material (DNA) into the egg. Eggs and spermatozoa contain the same amount of nuclear DNA despite their enormous difference in size. Since eggs contain many more mitochondria than spermatozoa, their cytoplasmic DNA content is much higher (1000 times or more). The size of the eggs varies greatly according to the species; but the enormous difference between the size of the egg of an ostrich and that of a sea urchin (which is barely visible to the naked eye) is mainly due to the amount of yolk (food material for

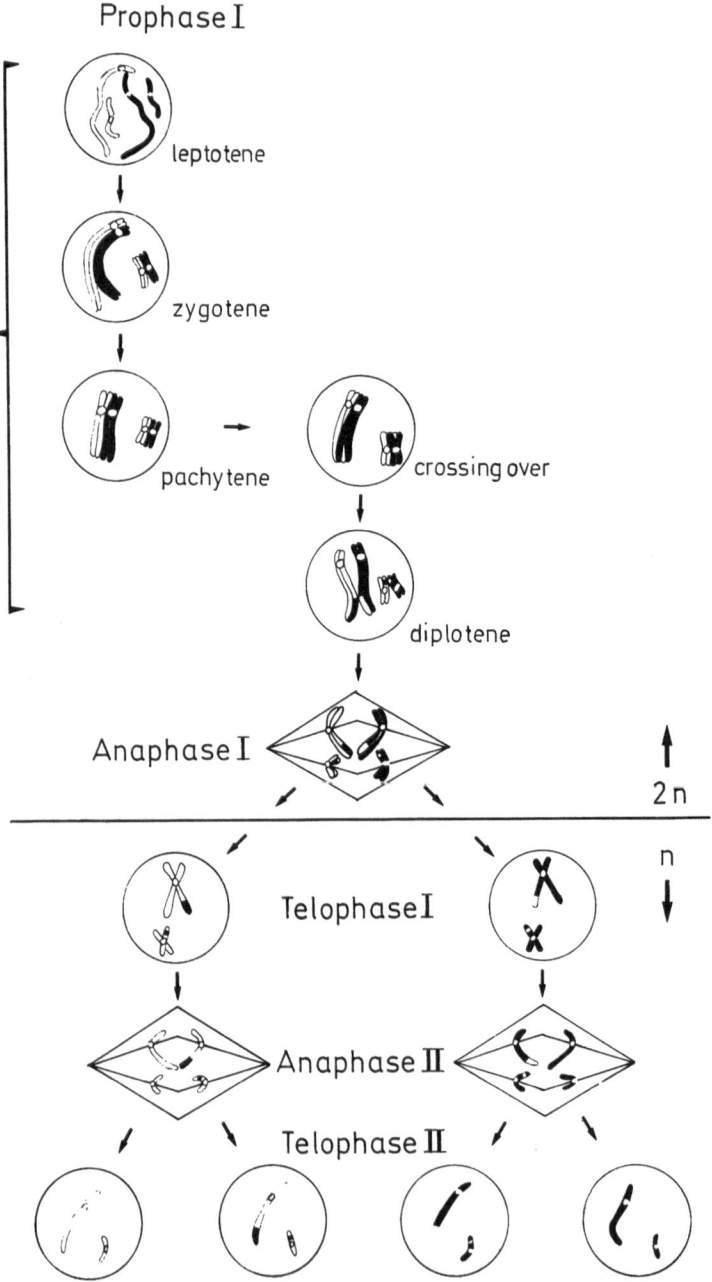

Fig. 11. The various stages of meiosis (nuclear changes)

the embryo) present. There is no correlation between the size of an egg and the amount of DNA which its nucleus contains. For instance, the egg of a frog and that of a newt are about the same size; yet the latter possesses at least six times more nuclear DNA than the former. However, the number of the genes necessary to produce a frog or a newt should be about the same. The large difference in nuclear DNA content between the two species (the so-called C-paradox) is due to the fact that the newt chromatin contains much more "selfish" DNA (repetitive sequences, introns, etc.) than its frog counterpart.

There is another major difference between eggs and spermatozoa. Eggs are the only cells which can give rise to a complete adult, i.e. they are *totipotent*. In contrast, spermatozoa are unable to even divide when they are cultured in media in which foodstuffs are generously provided, i.e. they are *nullipotent*. This is not due to differences in genes or nuclear DNA between the two gametes, but to the large differences in size and composition of their cytoplasms.

It should be added that oocytes and unfertilized eggs almost always display a distinct polarity. They have a light *animal* pole and a heavy, yolk-laden, *vegetal* pole.

Fertilization is immediately followed by *cleavage*. This is a period of quick and intense nuclear replication (mitotic divisions). The division of the egg cytoplasm is seldom perfectly equal, but the cleavage furrow usually starts from the animal pole and proceeds toward the vegetal one. There are several types of cleavage patterns, which are, by and large, related to the amount and distribution of the yolk (which constitutes an obstacle to the progress of the cleavage furrows). For instance, cleavage is only partial and limited to the aminal pole in the very yolky eggs of fishes and birds; it is superficial in those of the insects, where the yolk is accumulated in the center of the egg. Figures 12 and 13 depict cleavage in three species which have been often studied by molecular embryologists: the unequal (spiral) cleavage of the mollusk *Ilyanassa* and the complete and equal cleavages of the sea urchin and frog eggs. In these eggs, the cells (called *blastomeres*) surround a central cavity, the *blastocoele.*

In all animal species, cleavage leads to progressive *cellularization;* as cleavage proceeds, the size of the cells becomes smaller and smaller. Since the volume of the nuclei does not change greatly during cleavage, the ratio between the nuclear and cytoplasmic volumes (the *nucleocytoplasmic ratio*) steadily increases. The speed of cleavage, in different species, varies greatly and depends largely on the speed of DNA replication. For instance, a *Drosophila* egg replicates its DNA in less than 4 min; in contrast, DNA replication requires many hours in a mouse egg where cleavage into two cells is unusually slow and takes as long as 1 day.

Once cleavage is over, at the *blastula* stage, *gastrulation* sets in. This is primarily a period of extensive and complex cell movements, which cannot be described here. In vertebrates, these movements are much more active on the dorsal than on the ventral side. Figure 14 depicts schematically the gastrulation of an amphibian egg. The final result of gastrulation is the same in all tridermal animals. The gastrula is formed of three distinct sheets of cells, the *ectoblast,* the *mesoblast* and the *endoblast,* which surround an inner cavity, the *archenteron.*

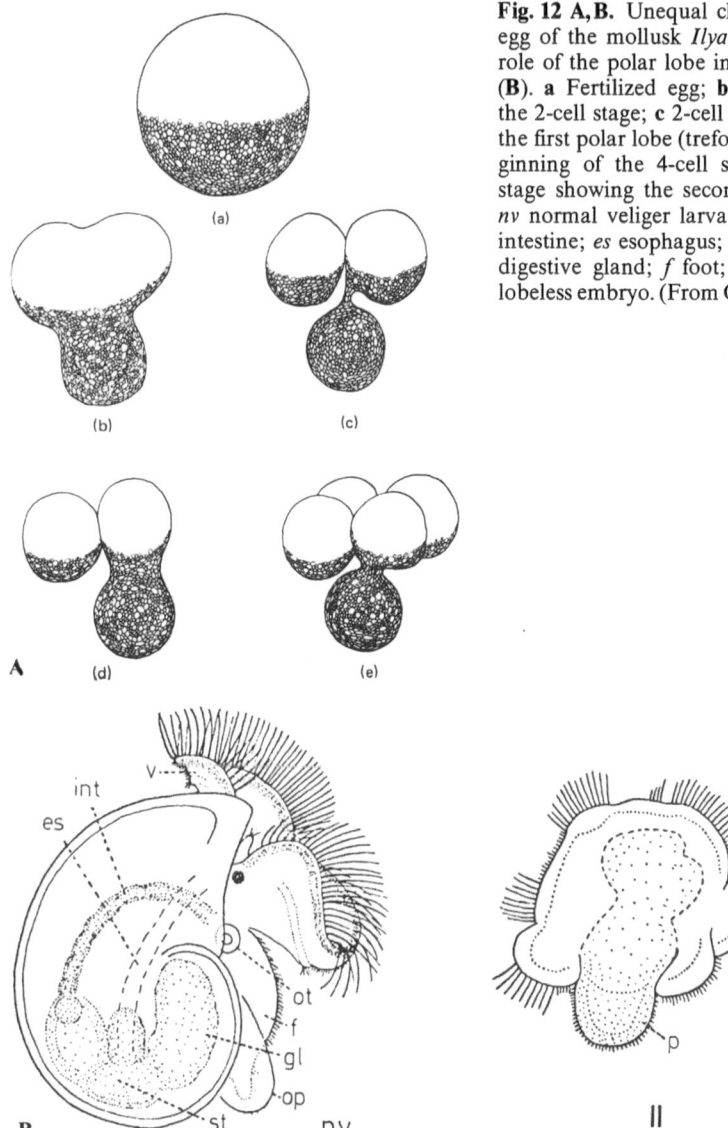

Fig. 12 A,B. Unequal cleavage of the egg of the mollusk *Ilyanassa* (**A**) and role of the polar lobe in development (**B**). **a** Fertilized egg; **b** beginning of the 2-cell stage; **c** 2-cell stage showing the first polar lobe (trefoil stage); **d** beginning of the 4-cell stage; **e** 4-cell stage showing the second polar lobe. *nv* normal veliger larva; *v* velum; *int* intestine; *es* esophagus; *st* stomach; *gl* digestive gland; *f* foot; *ot* otocyst; *ll* lobeless embryo. (From Clement 1952)

Later development is too variable to be described here; more will be said about it when we come to the discussion of special cases such as the sea urchin, the amphibian or the mouse embryo. It should be pointed out that, as a general rule, organ formation (organogenesis) precedes final cell differentiation. For instance, the nervous system of an amphibian embryo is well formed long before any nerve cell (neurone) has differentiated, morphologically, physiologically and biochemically.

Fig. 13 A, B. Complete and equal cleavage of a sea urchin (A) and a frog (B) egg. **a, g** Fertilized egg; **b, h** 2-cell stage; **c, i** 4-cell stage; **d, j** 8-cell stage; **e** 16-cell stage; **f, k** morula; **l** young blastula. Note the grey crescent in **j**

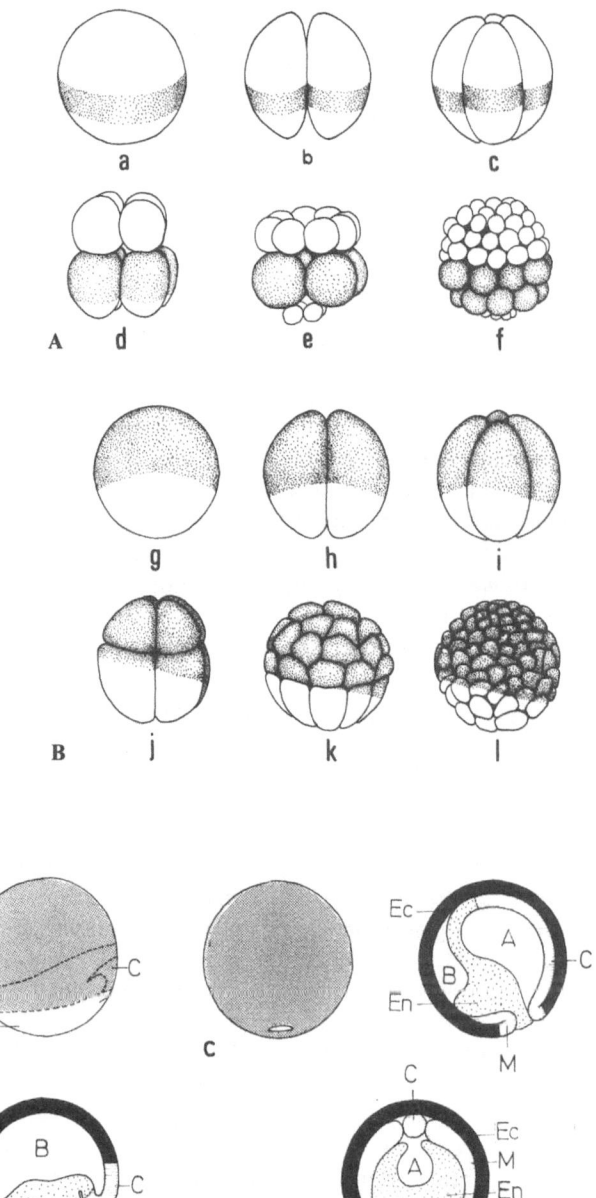

Fig. 14 a–d. Gastrulation in an amphibian embryo. **a** Early gastrula. *Left:* dorsal view; *right:* lateral view. **b** Middle gastrula. *Left:* dorsal view; *right:* section. **c** Late gastrula at the yolk plug stage. *Left:* dorsal view; *right:* vertical section. **d** Same stage as **c** transversal section. *A* archenteron; *B* blastocoele; *C* chordoblast; *Ec* ectoderm; *En* endoderm; *M* mesoderm

1.2 A Few Important Experimental Facts

It is customary, but not quite correct, to divide eggs into two different categories, on the basis of their reactions to experimental approaches such as destroying a part of the egg by pricking, transplanting or explanting groups of cells, centrifuging the egg, etc. There are *regulative* and *mosaic eggs.*

Embryonic regulation was discovered by Driesch long before the word regulation became popular among molecular biologists. He found that in sea urchin eggs, the separation of the first two blastomeres leads to the formation of two normal embryos (Fig. 15). If the blastomeres had not been separated, each of them would have produced only a half embryo. Therefore, the isolated blastomere has acquired higher potentialities for development than the normal one.

The fact that a part of an egg can give rise to a whole embryo seemed so surprising to Driesch that he based a neovitalistic philosophical theory on this experiment. Regulation would be due to the intervention of an *entelechy*, a principle which, by definition, cannot be studied experimentally. Today, embryologists are trying to understand the molecular basis of regulation without calling on entelechies and other vital forces for an explanation.

In *mosaic* eggs, like that of the mollusk *Ilyanassa*, removing part of the cytoplasm results in the formation of abnormal embryos, which have some parts missing. For instance, removing the *polar lobe* (which is purely cytoplasmic and never contains a nucleus) at the "trefoil stage" leads to a defective organism, in which the foot, eye and shell are either absent or abnormal (Fig. 12 B). Close analysis of operated mosaic eggs shows, however, that regulation is never completely lacking.

Many experiments made on amphibian eggs have confirmed the importance of the *cytoplasm*, at early stages of development, for morphogenesis. For instance, centrifugation of a recently fertilized frog egg leads to the formation of microcephalic embryos, which may lack brain and eyes. In contrast, if a fertilized frog egg is placed upside down (with its animal pole downward) and is slightly compressed, double embryos (siamese twins) are formed. Soon after fertilization, frog eggs show a particular pigmented area, they *grey crescent,* which indicates the future dorsal side of the embryo and the adult [Figs. 13 (2) and 54]. If one destroys it by pricking, very abnormal embryos, which are almost devoid of a nervous system, will form.

Another kind of cytoplasmic germinal localization, which is of major biological importance, can be seen in the eggs of a number of species. It is the *germ plasm,* which is required for the production of the reproductive organs (ovaries, testes) and cells (eggs, spermatozoa). In amphibian eggs, a material containing large granules, made of proteins and RNA, accumulates at the vegetal pole of the unfertilized egg; a morphologically, very similar material (called pole plasm) is localized at the posterior end of a number of insect eggs, including *Drosophila.* If these particular areas of the eggs are UV-irradiated, development is normal, except that the adults are sterile. It has been shown that in frog and *Drosophila* eggs, injection of the pole plasm (containing large ribonucleoprotein granules)

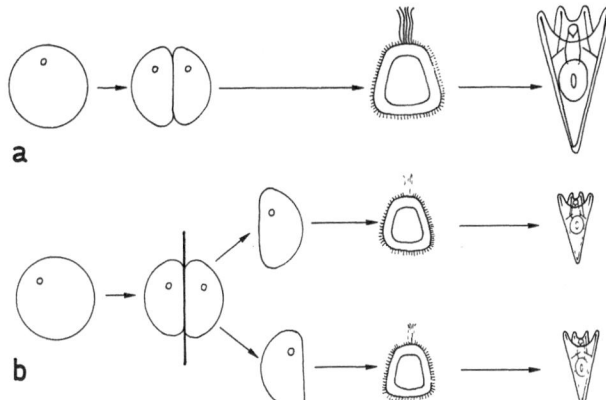

Fig. 15a,b. Regulation in sea urchin eggs (H. Driesch). **a** Normal development leading to the formation of a pluteus larva. **b** The separation of the two first blastomeres leads to the formation of two smaller, but normal plutei

into a UV-irradiated egg allows the formation of normal ovaries or testes. The nature of the active material is not known, but its strong UV-sensitivity indicates that nucleic acid is an essential factor. In other insects (*Cecidyomidae*), the nuclei which, during cleavage, reach the germ plasm, retain their chromosomes in the usual way during mitosis; in the other parts of the same eggs, a loss of chromosomes is observed. Thus, in these insects, only the nuclei, which will ultimately give rise to the gametes, retain their full chromosome complement; the somatic cells have a decreased number of chromosomes. If the germ plasm is disrupted by centrifugation, it forms scattered islets in these eggs; only the nuclei, which by accident lie in these islets, retain all their chromosomes. A similar phenomenon occurs in the Nematode *Ascaris*. It is *chromatin-diminution,* discovered by Boveri at the end of the nineteenth century. In *Ascaris* only the germ line cells retain intact chromosomes; in all the other cells, parts of the chromosomes are eliminated into the cytoplasm, where they are quickly degraded. Recent biochemical work has shown that the eliminated chromatin contains mainly repetitive DNA sequences. Apparently, these sequences are not needed for the somatic cells, but must be retained in the germ line cells which will produce the gametes. In all these cases, the behavior of the chromosomes is controlled by the chemical composition of the cytoplasm surrounding the nuclei.

As one can see, these experiments (and others that will be summarized later) show that many eggs are a *mosaic* of territories possessing different developmental fates and potentialities. These territories, which have been called *germinal localizations* by experimental embryologists early in this century, are different from the rest of the egg cytoplasm because they possess specific *cytoplasmic determinants* (as opposed to the nuclear determinants, i.e. the genes). Our limited knowledge of them will be presented in subsequent chapters.

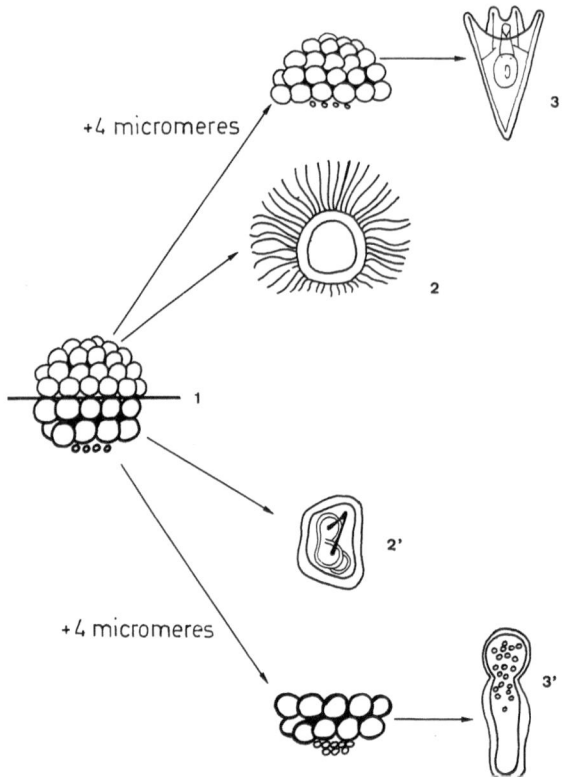

Fig. 16. Morphogenetic gradients in sea urchin eggs (S. Hörstadius). A morula is divided into an animal and a vegetal half (*1*). The former gives rise to an abnormal hyperciliated larva (*2*). The latter produces an abnormal pluteus (*2'*). If micromeres are grafted into the animal half, a normal pluteus is formed (*3*). If they are grafted into the vegetal half, exogastrulation occurs (*3'*)

The *respective roles of the nucleus and the cytoplasm* become more difficult to analyze when development proceeds further. Many experiments clearly show that the egg cannot be considered as a homogeneous sphere and that interactions between cells play an essential role in morphogenesis. We have seen the results of Driesch's experiments on sea urchin eggs at the 2-cell stage; if one repeats these experiments (meridian sections) during later stages of cleavage, regulation persists to a large extent. But if the section is an equatorial one (Fig. 16), the animal halves stop development at the blastula stage or form very abnormal larvae, which possess no gut. The vegetal halves, on the contrary, form such an excess of gut that the latter cannot invaginate (exogastrulation). If one grafts *micromeres* (which are the most vegetal cells) into isolated animal halves, the formation of an archenteron is induced and perfectly normal larvae (plutei) can be obtained. These experiments demonstrate the existence, in sea urchin eggs, of opposing *morphogenetic gradients* (Hörstadius). Certain substances, which are required for normal development, decrease progressively in concentration from the animal to the vegetal pole. Other substances have the opposite distribution with a maximal concentration at the vegetal pole.

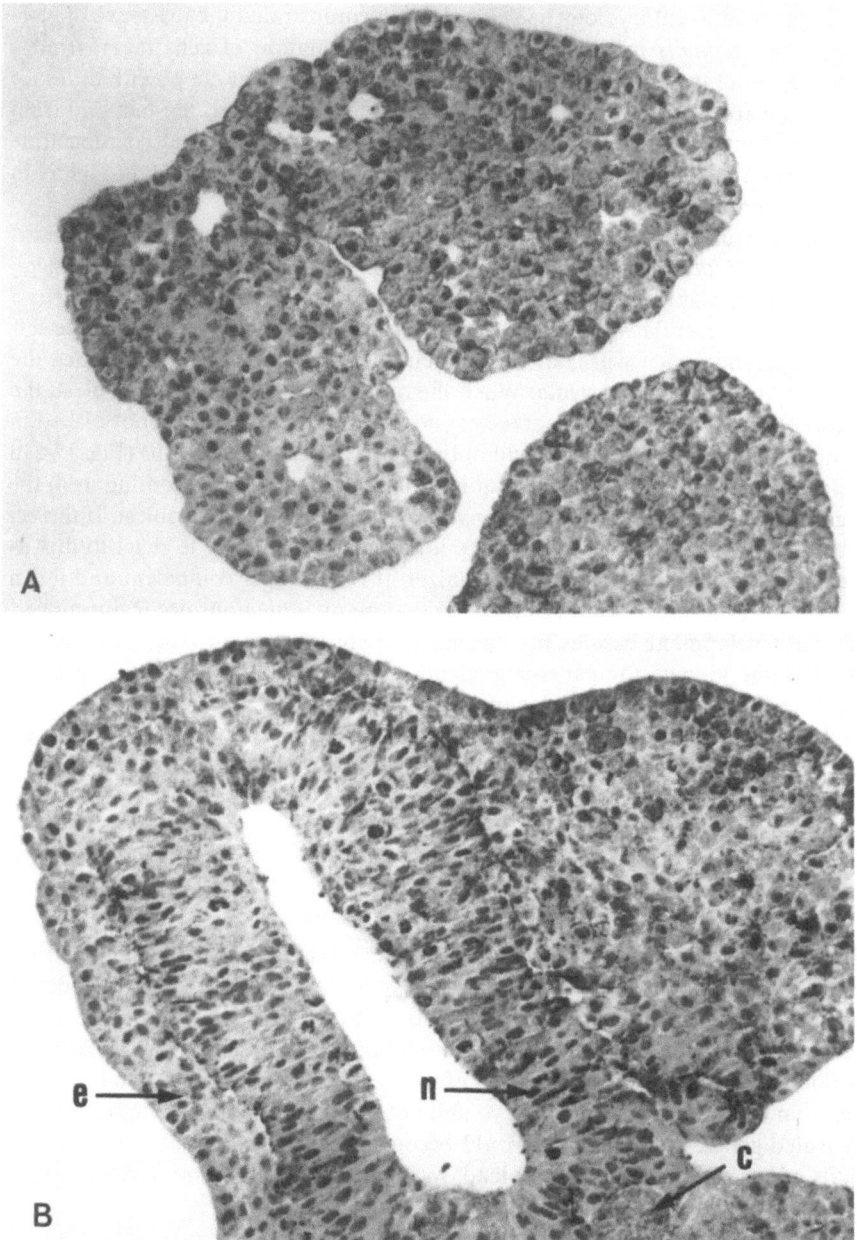

Fig. 17 A, B. Induction in an amphibian embryo. **A** Ectoblast isolated from an old blastula (axolotl) gives rise to ectoderm only. **B** Isolated ectoblast placed in contact with the dorsal lip of a young gastrula (chordoblast) gives rise to the nervous system and ectoderm. *c* chordoblast; *e* ectoderm; *n* nervous system. (Courtesy of Dr. R. Tencer)

Experimental embryology has also clearly demonstrated the existence of *morphogenetic gradients* in amphibian eggs. Transplantation of cells taken from a given region of the embryo into another area has shown that the potentialities for development tend to decrease gradually from the head toward the tail, and from the dorsal toward the ventral parts of the egg. Such experiments demonstrate that *dorsoventral* and *cephalocaudal* gradients are extremely important factors in the development of vertebrate embryos.

Another fundamental factor in vertebrate embryology is the existence of morphogenetic *inductions*. The best known case is that of Spemann's "organizer," which is responsible for the formation of the nervous system. The organizer is a region of the gastrula which corresponds roughly to the grey crescent of the fertilized egg and which will later on differentiate into chorda. This is, in fact, the dorsal lip of the young gastrula. When the organizer acts upon the ectoblast, the latter is transformed into the nervous system; when the ectoblast is not placed in contact with the chordoblast, it will only give rise to ectoderm (skin) (Fig. 17). In neural induction, two factors are required: the *inducing stimulus* coming from the organizer (chordoblast) and the *"competence"* of the reacting ectoblast. If the ectoblast is taken from an aged gastrula, it has lost the capacity to react to the inducing stimulus coming from the organizer. It is no longer competent and it can only form epidermis. It should be added that many inductions occur during vertebrate development besides the "primary" neural induction. The formation of the lens, the kidney, the pancreatic gland, etc., is due to inductive processes, in which cells from two different layers interact.

It is one of the main goals of molecular embryology to throw some light on the biochemical basis of the morphogenetic gradients and inductions discovered by experimental embryologists.

1.3 Genetic Theories of Morphogenesis

The great geneticist and embryologist T. H. Morgan realized that cytoplasmic heterogeneity, which is so important in the very early stages of embryonic development, might exercise an effect on genetic activity. He suggested, in 1934, that identical nuclei are distributed in a heterogeneous cytoplasm during the cleavage period which follows the fertilization of the egg. As a result, genes would become active in certain parts of the embryo and not in others. Under the impulse of the activated genes, the cytoplasm would become still more different in the various parts of the embryo. This would lead, by a cascade mechanism, to further gene activation and, ultimately, to cell differentiation.

Gurdon has demonstrated that the nuclei of differentiated cells present in the adult still contain the same genes as the fertilized egg and that they have *not* undergone *irreversible* differentiation during development. This very important fact has been shown by nuclear transfer experiments first done by Briggs and King (Fig. 18). Nuclei from tadpoles or adults have been injected into unfertilized eggs of the toad *Xenopus* after destruction by UV-radiation of the egg chro-

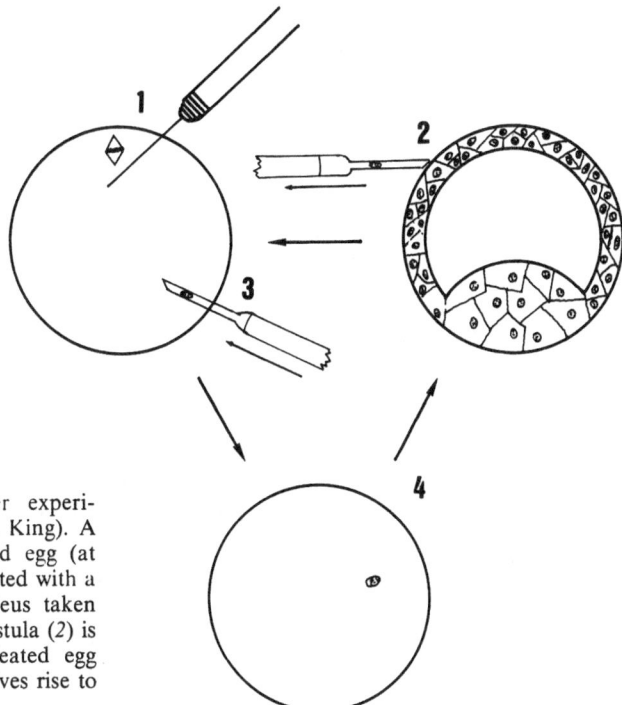

Fig. 18. Nuclear transfer experiment (R. Briggs and T. King). A *Rana pipiens* unfertilized egg (at metaphase II) is enucleated with a fine needle (*1*). A nucleus taken from a *Rana pipiens* blastula (*2*) is injected into the enucleated egg (*3*). The operated egg gives rise to a normal blastula (*4*)

mosomes. Perfectly normal larvae developed from a small percentage of the injected eggs (many of the grafted nuclei are injured and become highly abnormal after their transplantation into the egg cytoplasm). It is clear that if an egg, having received a nucleus from an adult, completely differentiated organ, can form normal embryonic tissues of every kind (nervous system, muscles, kidney, liver), then it must be assumed that the adult nucleus has retained many, if not all the potentialities of the nucleus of the fertilized egg.

Gurdon's experiments do not, of course, show that the *activity* of the genes is the same in the adult and in the just fertilized egg. For this reason, Morgan's theory remains one of the best we have to explain morphogenesis in embryos. Today, we think that the nuclear genes are *repressed* (which means that they are unable to synthesize mRNAs and that they cannot direct the synthesis of new proteins) during the very early cleavage stages of development. They would undergo *derepression* later; in other words, synthesize specific mRNA molecules which will allow the cytoplasm to build up new proteins. Morphogenesis would result from *differential gene activation*. The genes which direct the synthesis of a given protein (hemoglobin, for instance) would become active in a given area at a given stage of development.

Several theories have tried to explain how gene activity is controlled during embryonic differentiation. Most of them are derived from the famous Jacob and

Monod model which has so successfully explained gene regulation in bacteria: the activity of *structural genes* (encoding specific proteins) is controlled by *regulatory genes* producing activators or repressors.

There is no doubt that such mechanisms of positive and negative gene regulation operate in eukaryotes as well as in prokaryotes; but the organization of eukaryotic chromatin is so complex that more elaborate models must be developed, if we wish to explain, on a genetic basis, embryonic differentiation in animals. The most elaborate model is that of Britten and Davidson (1969) and Davidson and Britten (1979) which has two great merits: it takes into account the heterogeneity of the egg cytoplasm and, at the molecular level, it proposes a regulatory role for nuclear RNAs copied on middle repetitive sequences. This very elaborate model will not be presented here in view of its great complexity and the fact that it is as yet impossible to test experimentally.

What seems to be certain is that we should not look for a single control mechanism of gene activation during embryogenesis; there is no doubt that many mechanisms, localized at all steps between the gene and the encoded protein, interplay during development. This point was made very clearly by Brown (1981) in an excellent discussion of the control of genetic activity during morphogenesis.

A first possibility, which was suggested by Scarano already in 1967, is that embryonic development might be controlled by *changes in the DNA molecules* themselves (DNA modification). We shall soon see what these changes might be at the molecular level. Gurdon's experiments, showing that some adult nuclei are still totipotent after transplantation into enucleated unfertilized eggs, are an obvious stumbling block to this kind of theory. However, one cannot exclude the possibility that biochemical changes take place in adult nuclei when they rapidly replicate in young cytoplasm; such changes might bring "adult DNA" back to an "early embryonic DNA" state. This is only a hypothesis; but it could be tested experimentally with the now available techniques.

A convenient method for a cell to increase the production of a given protein (hemoglobin, for instance, in hematopoietic cells) is to increase the number of copies of the structural gene encoding that protein. This process, called *gene amplification,* has been proposed by Holtzer and colleagues (1972) as a major factor in cell differentiation. A few cases of gene amplification are known: this process occurs before the synthesis of the large ribosomal RNAs during amphibian oogenesis, as we shall see in the next chapter and it also occurs in the cells which surround *Drosophila* oocytes when they must synthesize large amounts of the chorion proteins. Gene amplification may also take place when cells become resistant to a few toxic agents (heavy metals, the anti-cancer drug methotrexate). But this remains the exception rather than the rule: it is now absolutely certain that the number of the globin genes does not increase when hematopoietic cells begin to synthesize hemoglobin.

Gene activation might also result from *gene rearrangements* due to translocations: genes initially separated in the genome might come in contact and as a result undergo new interactions. This occurs during lymphocyte differentiation, but again this seems to be an exception rather than the rule. We have seen that mobile genetic elements can be inserted in many sites of the genome in *Drosophila*. This may modify the eye color in this insect, but it is unlikely that insertion of mobile genetic elements into DNA plays an important role in early morphogenesis (for instance, in the formation of the eye itself).

More convincing is the proposition (Scarano 1969; Razin and Riggs 1980) that *DNA methylation and demethylation* might play an important role in embryonic differentiation. Many cells, including sea urchin eggs, possess DNA methylases, enzymes which add a methyl ($-CH_3$) group to some of the cytosine residues present in the DNA molecules. There are probably no DNA demethylases, but removal of the methyl groups occurs if DNA is replicated during several cycles of cell division in the absence of DNA methylation. There is no strong evidence so far for the view that DNA methylation increases during embryonic development; but there is increasing evidence that gene activity is often correlated with DNA *undermethylation*. In tissues where a specific gene is strongly expressed (the hemoglobin gene in hematopoietic cells, for instance), this gene and the adjacent sequences are undermethylated as compared to the corresponding DNA sequences in cells which do not synthesize hemoglobin. It is quite possible that in Gurdon's nuclear transplantation experiments, the DNA of the injected adult nuclei undergoes progressive demethylation during the repeated, fast cleavage cycles. Indirect evidence for the importance of DNA undermethylation in cell differentiation comes from the effects of *5-azacytidine,* an inhibitor of DNA methylation: treatment with this inhibitor has been reported to induce the differentiation of muscle cells and to reactivate the inactive X chromosome of mammalian females (see Chap. IX).

However, it seems probable that DNA undermethylation is only one of the many factors which are required for *chromatin activation* (review by Weisbrod 1982). This point has already been briefly discussed in the preceding chapter, in which we have seen that active chromatin displays a number of pecularities (high sensitivity to DNase digestion, differences in the content in various DNA binding proteins, etc.). To this list, one can add DNA undermethylation, but there is growing evidence that this process cannot be the only factor responsible for the genetic control of differentiation, e.g. in several cases, undermethylation is not correlated with gene activation.

The most important controls of genetic activity occur at the *transcriptional* and *post-transcriptional* levels: if a gene is not transcribed by an adequate RNA polymerase, it will remain silent; if the primary transcript of a gene is not properly converted into a mature mRNA molecule, transcription has been useless; if a mature mRNA molecule does not come out of the nucleus to bind to cytoplasmic ribosomes, the corresponding protein will not be expressed. There are thus a multiplicity of possible regulatory steps between the gene and its final product, a specific protein. An accident, at any one of these steps, will lead to the production of an abnormal (nonsense) protein or of no protein at all. This occurs in the *thalassemias;* these hereditary diseases, in which the hemoglobin is abnormal, may have a variety of molecular origins (deletion or mutation of a globin gene, abnormal processing of the primary gene product, etc.).

As we shall see when we study oocyte maturation and egg fertilization, a number of *translational* controls operate at the level of the cytoplasmic protein synthesizing machinery: efficiency of mRNA binding to ribosomes, stability in the cytoplasm of the mRNAs and of the proteins are very important regulatory factors. In fact, such translational control mechanisms may well be more important than gene transcription during the very early stages of development. However, sooner or later, gene transcription takes the upper hand and the major control mechanisms (transcriptional and post-transcriptional) then lie in the cell nuclei.

Two decades ago, it was proposed that the cytoplasm might contain self-reproducing specific particles (the hypothetical *plasmagenes*). It is true that a few cytoplasmic organelles can self-reproduce: the centrioles of the dividing cells, the cilia and flagella, the mitochondria of all eukaryotic cells and the chloroplasts of the green plant cells. However, there is no evidence for a specific role of these cell organelles in embryonic differentiation. Although both mitochondria and chloroplasts contain DNA and can therefore be the site of cytoplasmic, non-Mendelian mutations, it is now clear that they have only a restricted autonomy toward the nucleus. Many mitochondrial and chloroplastic enzymes are encoded by nuclear genes; these enzymes are synthesized on cytoplasmic ribosomes and move from the cytoplasm into the mitochondria or the chloroplasts. The information present in mitochondrial and chloroplastic DNAs is too restricted to allow full independence toward the nucleus of the organelles which possess their own DNAs. We have mentioned that the cytoplasm of mosaic eggs contains cytoplasmic determinants which are of fundamental importance for further development. However, there is no evidence that these determinants are endowed with genetic continuity, i.e. that they can self-reproduce as the purely hypothetical plasmagenes were supposed to do. These *cytoplasmic determinants,* as we shall see, are not the mitochondria. It seems more likely that they are accumulations of specific maternal mRNAs which had been synthesized during oogenesis. These mRNAs are presumably bound to proteins in the form of ribonucleoprotein particles. Therefore, for the time being, this remains a likely hypothesis, but not a hard fact.

CHAPTER IV

Gametogenesis and Maturation:
The Formation of Eggs and Spermatozoa

1 Oogenesis

The formation of the egg, in the ovary, is an extremely important period in on-togeny. It is much more than an accumulation of food material, which will be progressively utilized until the embryo becomes capable of feeding itself. It is al-so a period of intensive genetic activity, which leads to the synthesis of many im-portant macromolecules (RNA, proteins) which will be used at later stages of development only.

Amphibian oocytes, especially those of *Xenopus laevis,* will be the main topic of this chapter. As shown in Fig. 10, oocytes arise from oogonia by growth. Oogonia multiply and replicate their DNA; growth begins after a last doubling of DNA in oogonia. During the whole oogenesis, the oocytes have therefore a DNA content which is four times that of a haploid cell. The various stages of meiosis take place in very small oocytes, which are abundant in just metamor-phosed froglets. During the entire growth period of the occyte, the chromosomes remain in the diplotene (Fig. 11) stage of meiosis and there is no synthesis of chromosomal DNA. Growth is due to two different processes: endogenous syn-thesis of a variety of molecules (RNA, proteins, glycogen, lipids, etc.) and uptake of exogenous yolk proteins by endocytosis.

As shown in Fig. 19, one can distinguish three main stages in *Xenopus* oogen-esis: previtellogenic oocytes (0.1–0.3 mm in diameter) are transparent; vitello-genic oocytes are opaque and their diameters range between 0.3 and 1.2 mm; full-grown oocytes, which are ready for maturation, measure about 1.2 mm in di-ameter and display a distinct polarity: they have a pigmented animal pole and a white, yolk-laden, vegetal pole.

1.1 Cytoplasm

Figure 20 is an electron micrograph of the outer layers of a full-grown *Xenopus* oocyte. *Microvilli* greatly increase the surface of the plasma membrane; their axis is made of actin microfilaments associated with other proteins (villin in particu-lar). They are not contractile, but depolymerization of the actin microfilaments induces their retraction when oogenesis has come to an end. The microvilli pro-ject into a protective membrane, the *chorion* or *vitelline membrane* and establish

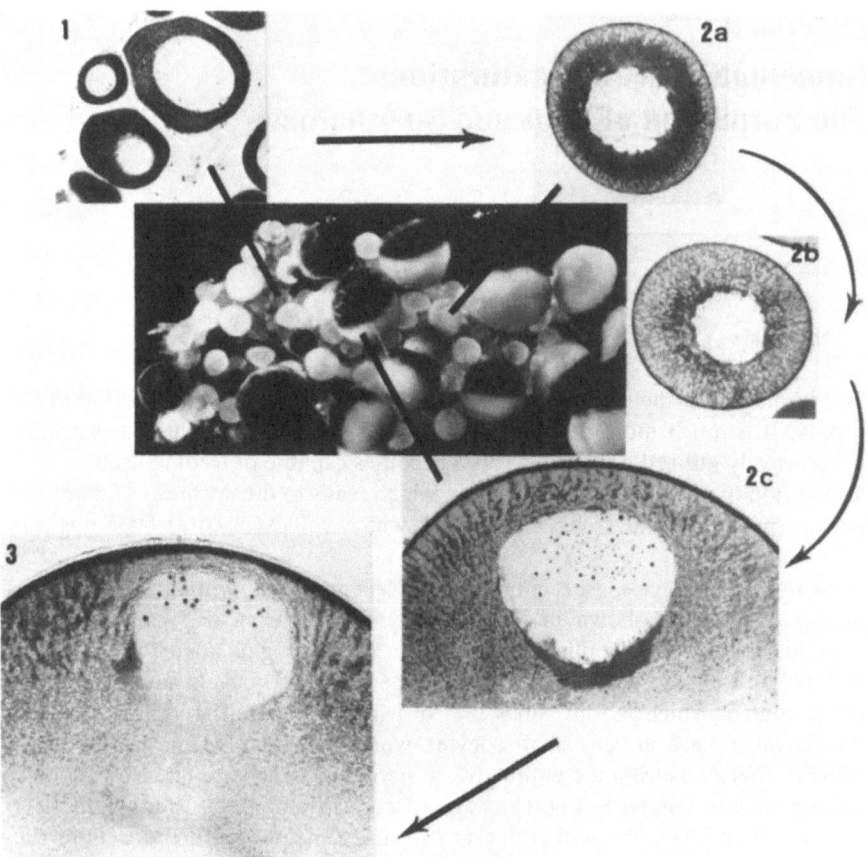

Fig. 19. Oogenesis of *Xenopus laevis*. *Center:* Fragment of ovary showing oocytes of various sizes; the full-grown oocytes have a pigmented animal pole and a white vegetal pole. *1* Section through small previtellogenic oocytes; *2a,b* two successive stages of vitellogenesis; *2c* full-grown oocyte showing, under the germinal vesicle, which contains many nucleoli, a cap of RNA-rich material. *3* Beginning of maturation: the nuclear membrane has broken down at its basal end, the nucleus with its nucleoli has moved toward the animal pole. The *bars* correlate the sections and the living oocytes at the various stages of oogenesis (Brachet 1979)

contacts with the surrounding layer of follicle cells (FC). Under the plasma membrane, in the so-called *cortex,* many *cortical granules* and pigment granules can be seen. The cortical granules contain glycoproteins and play, as we shall see, an important role in fertilization. The pigment granules are an accumulation of waste products from protein metabolism.

The role of the microvilli is to increase the penetration of blood proteins into the oocyte by *endocytosis.* Selective uptake of the yolk proteins is facilitated by the fact that the oocytic plasma membrane possesses specific *receptors* for the

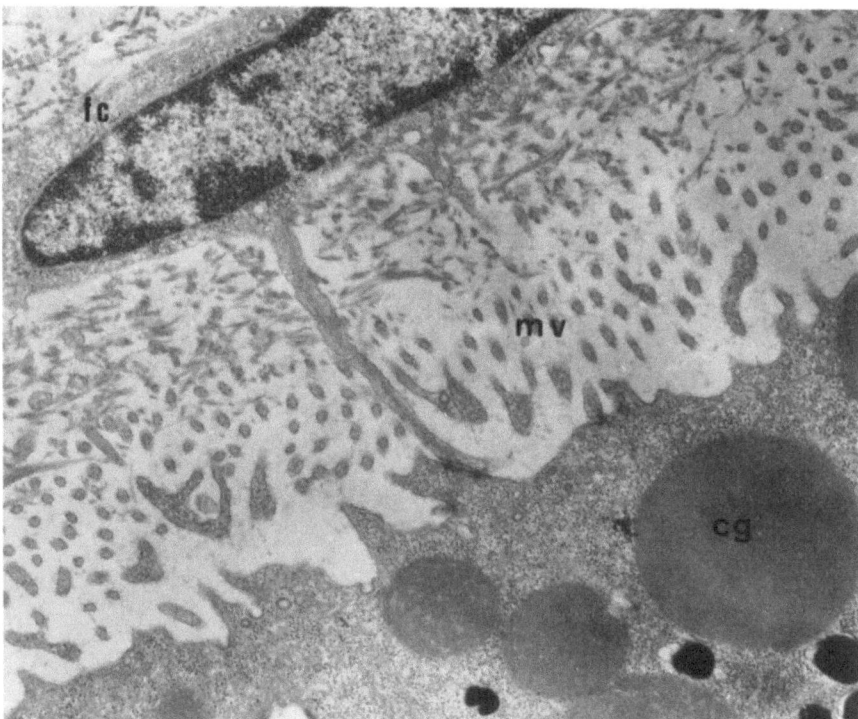

Fig. 20. Electron micrograph of the surface of a *Xenopus* oocyte. *fc* follicle cell; *cg* cortical granule; *mv* microvilli of the oocyte. (Courtesy of Dr. G. Steinert)

precursor of the yolk proteins. This precursor, called *vitellogenin,* is synthesized, after hormonal stimulation, by the liver of the female, released in the bloodstream and finally taken up into the oocyte by endocytosis. Vitellogenin is a complex of two phosphoproteins, *phosvitin* and *lipovitellin;* this phosphoprotein complex (which bears some similarities with the casein of milk) accumulates, in crystalline form, in the *yolk platelets.* The uptake of vitellogenin is the major factor in oocyte growth. It has indeed been shown that if full-grown oocytes are dissected from the ovary and cultured in the presence of vitellogenin and insulin, their volume doubles within 2–3 weeks.

The yolk platelets are larger and more abundant at the vegetal pole of the oocytes. In contrast, ribosomes, glycogen particles and lipid droplets gradually decrease in number from the animal to the vegetal pole. There is, thus, in full-grown oocytes, a very distinct polarity gradient (Fig. 21), the establishment of which does not result from simple factors such as gravity or blood supply. One of the factors responsible for the building up of the animal-vegetal polarity gradient is a transcellular electric current due to the movement of ions through the oocyte.

There is an enormous increase in the number of *ribosomes* during vitellogenesis. In addition to their distribution along the polarity gradient, a local accumu-

PA

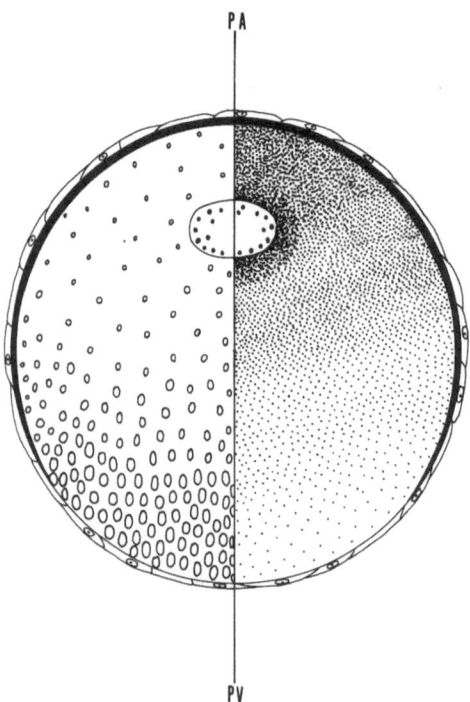

PV

Fig. 21. Polarity gradients in an amphibian oocyte surrounded by follicle cells. *PA* animal pole; *PV* vegetal pole. *Left half:* distribution of the yolk platelets. *Right half:* Distribution of the ribosomes; note their accumulation around the nucleus (germinal vesicle). In the cortex, gradient distribution of the pigment granules (melanosomes) (Brachet 1979)

lation of ribosomes around the large nucleus (called the *germinal vesicle*) of the oocyte can be seen. Most of the ribosomes, in full-grown oocytes, are free. However, others are attached to membranes to form a discrete endoplasmic reticulum or are in the form of polysomes.

Previtellogenic oocytes have very few ribosomes and synthesize only very small amounts of high molecular weight rRNAs (28 S and 18 S). However, they synthesize very actively the small 5 S rRNA, as well as several tRNAs (C. Thomas). The neosynthesized 5 S and tRNAs are packed together in combination with proteins into *ribonucleoprotein particles* smaller than the ribosomes. In these particles, 5 S RNA is strongly bound to a protein which, as we shall see, plays an important role in the control of 5 S RNA synthesis. During vitellogenesis, synthesis of 5 S RNA continues, but the oocyte now concentrates on the synthesis of the larger 28 S and 18 S rRNAs. This allows the accumulation of a large store of ribosomes, much larger than is needed for the synthesis of proteins in the growing oocyte.

Due to the virtual absence of ribosomes in previtellogenic oocytes, protein synthesis is weak at this early stage of oogenesis. It increases during vitellogenesis, although in full-grown oocytes, it decreases in intensity along the animal-vegetal gradient of ribosome distribution as shown in Fig. 21.

Amphibian oocytes also synthesize a variety of *mRNAs* amounting to 3% to 5% of total RNA. It was believed until recently that this synthesis takes place during vitellogenesis, but recent work has shown that the mRNAs are synthesized mainly, if not exclusively, by the previtellogenic oocytes. They are then stored, in an almost inactive form, during the whole vitellogenic period. This is true even for the histone mRNAs, which are apparently useless in the growing oocytes, in which chromosomal DNA is not replicated. The mRNAs synthesized by the very young oocytes and kept in reserve during vitellogenesis form a very heterogeneous population. There are about 20,000 different mRNA species, theoretically capable of directing the synthesis of as many proteins.

In conclusion, the onset of vitellogenesis marks a sharp turning point in oogenesis. Synthesis of a large mRNA population stops and production of the large ribosomal RNAs begins, continuing during the whole oogenesis. The synthesis of the small RNAs (5 S RNA, tRNAs) is not suppressed by vitellogenesis. Whether these changes in the pattern of RNA synthesis are due directly to the uptake of vitellogenin and the appearance of yolk platelets is not known.

Elegant experiments by Gurdon and colleagues have proved that in full-grown *Xenopus* oocytes, the great majority (about 98%) of the ribosomes are *not* engaged in protein synthesis. If one microinjects a specific mRNA into *Xenopus* oocytes, coding for rabbit hemoglobin for instance, then the injected oocytes synthesize *rabbit hemoglobin* for many days. This shows that the great majority of the oocyte ribosomes are "unprogrammed", i.e. they are free to accept and bind any kind of RNA which is presented to them. Since these pioneer experiments, *Xenopus* oocytes have become an ideal system for testing whether an RNA isolated from any kind of tissue is a messenger. If so, synthesis of the encoded protein takes place in injected oocytes. Recent work has shown that the oocyte is even capable of excreting the foreign proteins if they are normally secreted by secretory cells. In this case, synthesis of the secretory protein, after injection of its mRNA, takes place on ribosomes attached to the oocyte endoplasmic reticulum membrane. Secretion of a protein, in all kinds of cells, requires the addition of sugar residues to the protein (glycosylation). This glycosylation step is accomplished as efficiently and faithfully in the oocyte as in secretory cells. Finally, the *Xenopus* oocyte system has allowed us to understand one of the pecularities of the mRNA molecules. Most of the mRNAs (but not all of them) possess at the 3' end of their molecule a *polyadenylic* (poly A) tract, which is a monotonous sequence made of a variable number of adenylic acid residues. If this (poly A) "tail" is enzymatically removed prior to injection into the oocyte, the stability of the injected mRNA is greatly reduced. Thus, synthesis of the encoded protein stops after a few hours instead of several days. Messengers which have no (poly A) tail (those encoding the histones, for example) have a very limited stability after injection into *Xenopus* oocytes. However, artificial addition of a (poly A) tail to histone mRNAs increases their longevity (G. Huez and G. Marbaix).

It should be added that the oocyte synthesizes many kinds of *enzymes* during oogenesis. It is interesting to note that some of these enzymes are apparently useless for the oocyte itself and serve only at a much later stage of development. This is particularly striking in the case of an enzyme involved in the synthesis of collagen, a protein which is not detectable in the amphibian egg before the gastrula and even the neurula stage. The enzyme is thus present in the oocyte long before its substrate is synthesized.

The oocyte foresees its future; it builds up large reserves of all kinds of RNAs, ribosomes, enzymes, food materials, in fact, everything it will need to survive the hardships of embryonic life, in a difficult world outside the maternal organism.

The egg needs a large supply of energy for the synthesis of new macromolecules and for the building up of structures; it is therefore not surprising to find that it contains a very large number of *mitochondria,* probably many more than are actually needed for oogenesis itself.

These mitochondria contain, besides the usual respiratory enzymes, molecules of circular DNA, which are about 5 μm long and have a molecular weight of 10 million daltons. This mitochondrial DNA synthesizes the mitochondrial rRNAs, which enter into the composition of the 55 S mitochondrial ribosomes; the latter use specific mitochondrial tRNAs for the synthesis of proteins. As pointed out before, the information present in mitochondrial DNA cannot allow the synthesis of a very large number of proteins.

Mitochondrial DNA can replicate independently of chromosomal DNA synthesis, thus allowing a huge increase in the number of mitochondria during oogenesis. This increase occurs mainly before vitellogenesis, e.g. out of a total of 16–17 rounds of mitochondrial DNA replication during oogenesis, 12 take place before the onset of vitellogenesis. Thus, a small previtellogenic oocyte already possesses 500,000 mitochondria. Such small oocytes also possess, as we have seen, more than 50% of the 5 S RNA and probably all of the mRNAs found in full-grown oocytes.

Mitochondria are also present in the spermatozoa. This leads to an interesting genetic problem. One would like to know whether new mitochondria are formed by replication of maternal (egg) or paternal (spermatozoon) organelles during embryonic development. In fact, an answer to this question has been given by experiments in which two different, but closely related species of the toad *Xenopus (X. laevis and X. borealis)* were crossed; the two parental species were selected because the properties of their mitochondrial DNAs are so different that they can easily be distinguished by physical measurements. The experiments have clearly shown that the mitochondrial DNA formed by the embryo is of the maternal type. This excludes the possibility that sperm mitochondrial DNA might play a major genetic role in embryonic development.

1.2 The Nucleus (Germinal Vesicle: GV)

The main constituents of the large oocyte nucleus are the following: the nuclear membrane (or envelope), the chromosomes, the nucleoli and the nuclear sap. They will be discussed in that order.

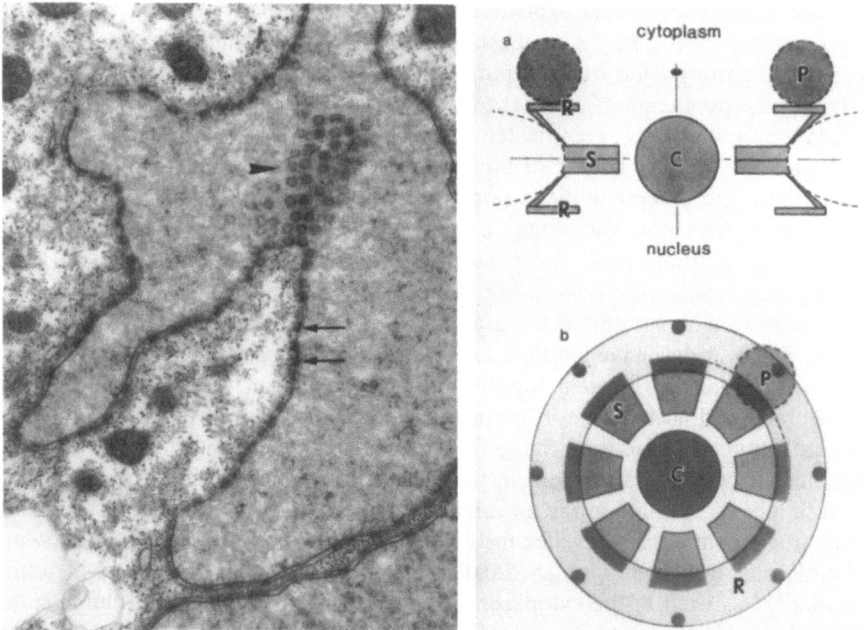

Fig. 22. Nuclear pores. *Left:* nuclear membrane of a *Xenopus* oocyte under the electron microscope; the micrograph shows cross-sections (*arrows*) and tangential sections (*arrowhead*) of nuclear pores. (Courtesy of Dr. G. Steinert.) *Right:* Diagram of the nuclear pore complex in a central cross-section (*a*) and in projection down the octad axis. (*b*) *C* central plug; *S* spokes; *R* rings; *P* ribosomes; *dashed line* nuclear membrane (Unwin and Milligan 1982; reproduced from *The Journal of Cell Biology,* 1982, 93: 63–75 by copyright permission of the Rockefeller University Press)

1.2.1 The Nuclear Membrane

This is made of two layers (outer and inner layers), as in all nuclei. However, it is thicker than in most cells. It is possible to isolate, by dissection, the germinal vesicle of an oocyte and, if one squashes it, one can easily see the remnants of the nuclear membrane under the light microscope. When examined under the electron microscope (Fig. 22), one is at once impressed by the regularity and large size of the *nuclear pores.* Closer analysis has shown that these pores are, in fact, very complicated structures, which have been called *nuclear pore complexes.* They have an octogonal shape, are reinforced by a system of very thin filaments and are obturated by a central granule, which possibly contains RNA. The diameter of the pores (60 nm) remains constant throughout oogenesis, but their number greatly increases. The nuclear pore complexes are held together by an interporous material, called the *lamina.* Pore-lamina complexes have been found in the nuclear membranes of all cells, e.g. in *Xenopus* oocytes, the lamina is made of a single protein, which is continuously synthesized and phosphorylated.

The nuclear membrane constitutes the borderline between the nucleus and the cytoplasm. Thus, RNAs synthesized in the nucleus continuously flow into the cytoplasm through the nuclear pores. Conversely, proteins synthesized in the cytoplasm cross the pores in order to reach the nucleus. The intensity of this traffic, in *Xenopus* oocytes, is indicated by the following figures: three molecules of 28S and 18S rRNAs go out of the nucleus, every minute, through each pore; since there are several millions pores, it can be calculated that about 300,000 rRNA molecules move out of the nucleus into the cytoplasm every second.

Thanks to its large size, the *Xenopus* oocyte is an ideal material for the study of *nuclear membrane permeability*. It is easy to inject substances of known molecular weights into the cytoplasm and to see whether they move into the nucleus. These studies have led to unexpected results and conclusions, e.g. whether or not a radioactive protein injected into the cytoplasm will cross the nuclear membrane and accumulate in the nucleus is independent of its size (molecular weight), shape or electrical charge. But proteins which are normally located in the nucleus very quickly return to the nucleus after injection into the cytoplasm and finally accumulate in the germinal vesicle (*karyophilic* proteins). This is not only true for small proteins, like the histones, but also for nuclear proteins having a molecular weight as high as 120,000. On the other hand, proteins which are normally localized in the cytoplasm of the oocyte do not move into the nucleus after their injection into the cytoplasm. This suggested that there is a selective permeability of the nuclear membrane for nuclear (karyophilic) proteins. However, recent experiments have shown that the selective uptake of karyophilic proteins into the nucleus remains unaltered when a small hole has been bored through the nuclear membrane with a fine needle. This leads to the conclusion that there is no nuclear membrane selective permeability. Thus, the selective uptake of the karyophilic proteins into the nucleus is probably due to their specific binding to nuclear constituents, not to a selection by the nuclear membrane. The transport of RNAs between the oocyte nucleus and its cytoplasm also raises interesting problems. For instance, the ribonucleoprotein particles which carry the rRNAs and the mRNAs from the nucleus to the cytoplasm are too large to cross the pores (the molecular weight of 28S rRNA is higher than 1 million daltons). Therefore, it is likely that these particles and even the RNAs, which they contain, undergo biochemical and conformational changes at the time they cross the nuclear membrane. Recently, it has been shown that if *small nuclear RNAs* (called snRNAs) are injected into the oocyte cytoplasm, they quickly move into the nucleus where they accumulate. In similar experiments, 5S RNA was found to accumulate in the nucleoli. But other small RNA molecules, the tRNAs for instance, remain in the cytoplasm after injection. The reasons for the different behaviour, in such experiments, of RNA molecules which have very similar molecular weights remain to be elucidated.

1.2.2 Chromosomes

In extremely young oocytes, which can be found in tiny *Xenopus* toads which have just undergone metamorphosis, the chromosomes are still at an early stage of meiosis (pachytene). Two distinct parts, both containing DNA, can be seen in the chromatin: a condensed "cap" made of rather homogeneous material, and the thick *chromosomes* which are closely paired at that stage (Fig. 23). The cap is involved in the formation of the nucleoli, as we shall see when we discuss these organelles.

When the oocyte begins growing during the period of vitellogenesis, the chromosomes expand considerably and take on a very peculiar shape. They are now the *lampbrush chromosomes* (Fig. 24), which are formed by a row of DNA-rich granules, the *chromomeres,* from which thin expansions, the lateral *loops,* project into the nuclear sap. The lampbrush chromosomes are at the diplotene stage of meiosis. As shown in Fig. 24 A, they undergo local pairing and are attached together at the so-called *chiasmata.* The "lampbrush"[1] aspect is due to the presence of the lateral loops, which are very easy to see in amphibian oocytes and are useful for cytogenetic analyses. It can occur that a loop is missing on one side of the lampbrush chromosomes and is present on the other; this is due to the lack of activity on one of the two allelic genes, in the heterozygous condition. It should be added that many cytologists think that the lampbrush structure is universal for all chromosomes, but greatly amplified and exaggerated in the growing amphibian oocyte.

Figure 24 B represents the structure of the lampbrush chromosomes in the schematic way proposed by Gall and Callan. According to this scheme, each chromosome is composed of two homologous filaments (*chromatids*), which are closely paired. Each chromatid is made of a single, giant (several centimeters long) fiber of double-stranded DNA; the chromomeres are regions where the DNA fiber is repeatedly coiled and condensed into microscopically visible granules. The axes of the loops are made of the same, but uncoiled, DNA fiber. The DNA present in the loops is transcribed into RNA (presumably mRNA) in a sequential way. The enzyme RNA polymerase wanders along the loop, starting at one end and stopping at the other. The arguments presented in favor of the model shown in Fig. 24 B are quite strong. When the lampbrush chromosomes are treated with enzyme deoxyribonuclease (DNase), which breaks DNA up into small pieces, the structure of the whole chromosome (chromomeres and loops) collapses and rapidly vanishes. Autoradiography (a photographic procedure, which enables one to see radioactive materials under the microscope) shows that RNA synthesis occurs in a sequential way from one end of the loop to the other.

---

[1] No one uses "lampbrushes" anymore. They were mops used for cleaning petroleum lamps before the electric light came into general use; natural selection has eliminated lampbrushes. The senior author remembers using them in the French countryside during boyhood. Today, they might be valuable as antiques.

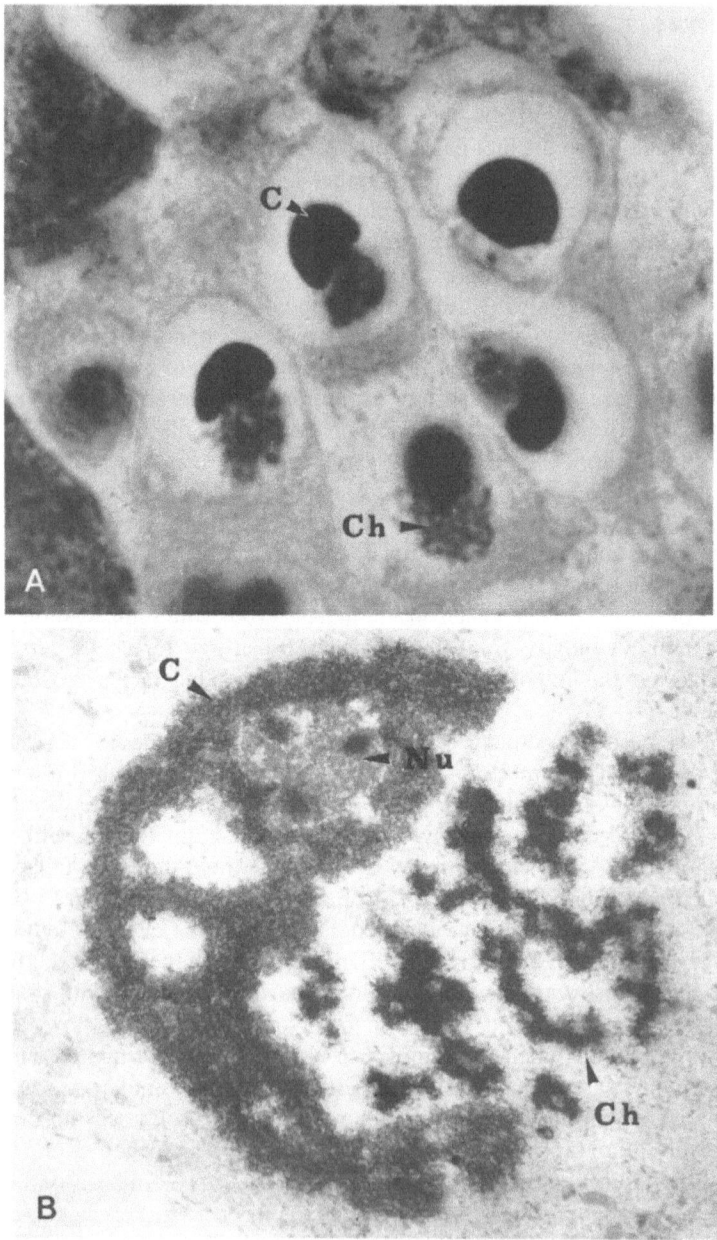

Fig. 23 A, B. Young oocytes of *Xenopus* at the pachytene stage. **A** Under the light microscope (A. Ficq). **B** With electron microscope (P. Van Gansen). *C* cap; *Ch* chromosomes; *Nu* nucleoli

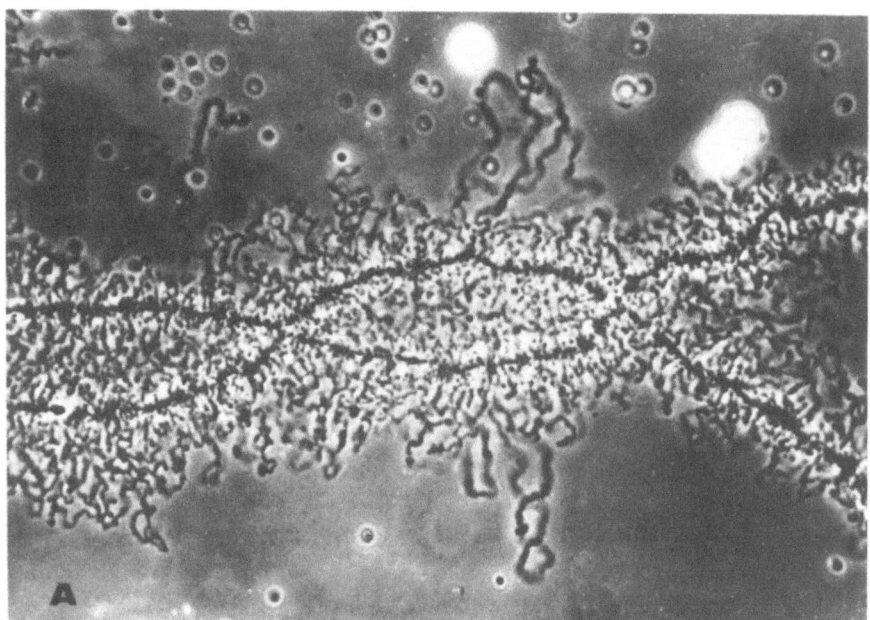

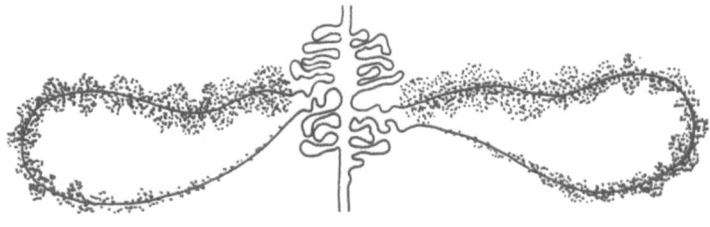

Fig 24 A, B. Part of a lampbrush chromosome isolated from a *Triturus* oocyte, observed with phase contrast optics (**A**) (courtesy of Dr. H. G. Callan) and diagram of the structure of a loop pair (**B**) (J. G. Gall and H. G. Callan): two DNA molecules are folded to form a *chromomere* from which a loop extends in the nuclear sap; it is the site of RNA synthesis (the small *black dots* are RNA molecules)

It was first believed that each of the 10,000 chromomeres corresponds to a single gene. Later, it was thought that each loop represents a single gene. However, the development of two powerful techniques has shown that the situation is more complicated. The first technique is *in situ hybridization* on whole chromosomes (which allows the detection of specific DNA sequences) and the second is chromosome *spreading* (chromosomes are spread on grids and examined under the electron microscope). Thus, the visualization of the so-called *transcription units* (see Fig. 27) is possible. We can now see the RNA chains growing during

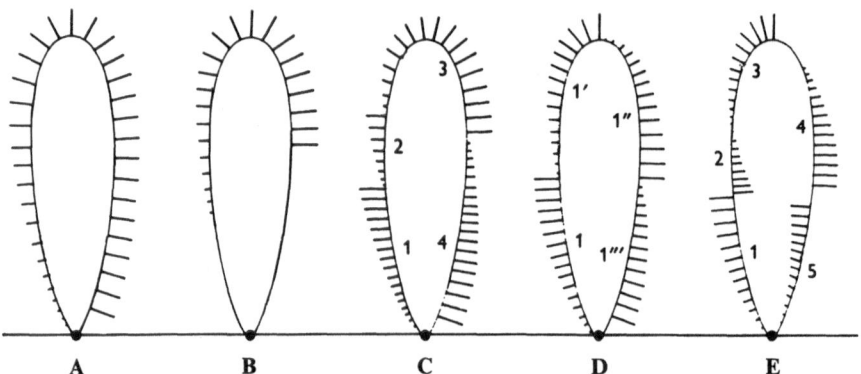

Fig. 25 A–E. Various alternatives for arrangements of transcription units with individual loops of *Xenopus laevis* lampbrush chromosomes. The numbers *1–1'''* denote units of equal, *1–5* of different lengths (Scheer et al. 1976)

the transcription of a single gene. These studies have shown that the loops contain both unique and repeated sequences. Curiously, only 0.2% of the lampbrush chromosome DNA codes for mRNAs and only 30% of their unique sequences are transcribed. The *primary transcriptional* products of the loops are very large (10–30 μm long, 5 million daltons) RNA molecules which associate immediately with proteins to form nuclear ribonucleoproteins. Further work has shown that several transcription units can be found on the same loop (Fig. 25), therefore, a loop is not a single gene, but a set of genes. In addition, it has been proven beyond doubt that highly repetitive DNA sequences are transcribed into RNA molecules in many loops of the lampbrush chromosomes and that many of these transcripts are transferred to the cytoplasm. If we recall that the whole mRNA population of the oocyte (about 20,000 different species) is synthesized in pre-vitellogenic oocytes (where the lampbrush structure is not yet very apparent), we are faced with a puzzling question: what is the role of the RNAs which are transcribed on the loops and are continuously supplied to the cytoplasm during months of vitellogenesis? We do not yet know the answer to this question, despite years of work in many laboratories.

At the very end of oogenesis, the lampbrush chromosomes progressively undergo condensation. They are then lying side by side in the center of the germinal vesicle; at the same time, their activity for RNA synthesis decreases. There is no doubt that there is a close correlation between lampbrush structure and activity of RNA synthesis: inhibitors of RNA synthesis (actinomycin D, antibodies against RNA polymerase II, etc.) produce the collapse of the loops and the contraction of the lampbrush chromosomes. Transcription takes place, as we have seen, only in the extended loops; the condensed chromomeres, in contrast, are not transcribed.

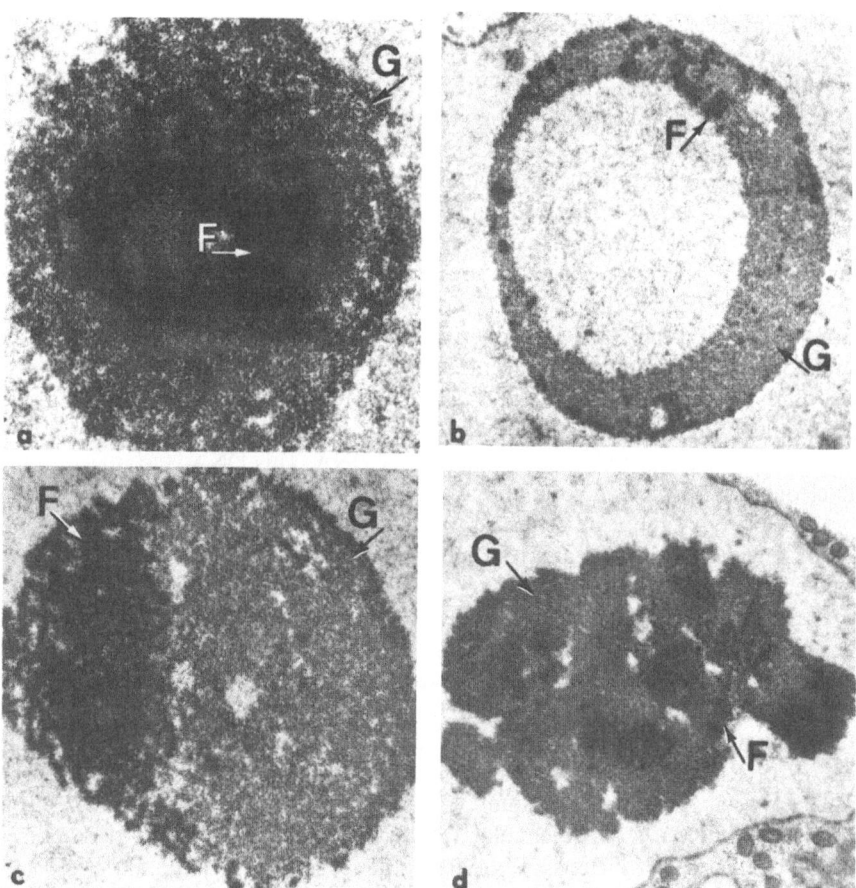

Fig. 26 a–d. Ultrastructure of nucleoli in the oocytes of four different amphibian species. **a** *Xenopus;* **b** *Triturus;* **c** *Bufo;* **d** *Pleurodeles. G* granular constituent; *F* fibrillar constituent. (Courtesy of Prof. P. Van Gansen)

1.2.3 Nucleoli

The number of nucleoli varies considerably among the oocytes of different species. While the oocytes of most invertebrates contain one single big nucleolus, the large oocytes of most vertebrates have hundreds of nucleoli of various sizes. There are more than 1000 large nucleoli in the germinal vesicle of the amphibian oocyte.

The *ulstrastructure* of the oocyte nucleolus is of the classical type (Fig. 26). Each is made up of a central fibrillar core and a granular cortex. However, the young previtellogenic oocytes of *Xenopus* which, as we have seen, do not synthesize the large 28 S and 18 S rRNAs, lack the granular part and are purely fibrillar.

Cytochemistry shows that the granular part is made of RNA-containing particles, which are the precursors of the cytoplasmic ribosomes.

The large, single nucleolus of the starfish oocytes can be isolated and analyzed by biochemical methods. It contains about 7% *RNA,* which like rRNA, is rich in guanine (G) and cytosine (C). Nucleolar RNA has, however, a still larger molecular weight (around 3 million daltons) than the rRNAs. This large molecule is the precursor of the 28S and 18S rRNAs of the cytoplasmic ribosomes. The rest of the nucleolus is made of proteins. They are mainly phosphoproteins and basic proteins; some of which are precursors of the ribosomal proteins. It is highly probable that the nucleolar proteins are entirely of cytoplasmic origin. Among these proteins are a number of enzymes involved in RNA synthesis or breakdown; of particular importance is the very active nucleolar RNA polymerase I, which is not sensitive to *α-amanitin* and which is specialized in the synthesis of the nucleolar and ribosomal RNAs.

This enzyme is inactive in the absence of DNA. This DNA is localized in the nucleolus itself, where it forms the so-called *nucleolar organizer.* The nucleolar organizer DNA cannot be observed with the classical cytochemical methods for DNA detection unless special conditions which induce condensation of the nucleolar organizer are used.

Miller has succeeded in isolating the nucleolar organizers from amphibian germinal vesicles. They can be seen as ring-shaped structures under the light microscope. Using the electron microscope, Miller obtained an extraordinary photograph of a "gene in full action" (Fig. 27) for the first time. The nucleolar organizer is made of a DNA axis which, at regular intervals, gives rise to growing chains of RNA. The RNA polymerase molecules can be seen in the electron micrographs as small dots all along the DNA molecule. The pictures clearly show that only some parts of the circular DNA are transcribed and give rise to polynucleotide chains of nucleolar RNA. The other regions, which remain genetically silent and are not transcribed by the polymerase, are called the *spacers.* While the transcription units (the "Christmas trees" of Fig. 27) have a constant length and the same base composition, the spacers are much more variable, e.g. their length may vary as much as 1.8 to 5.5 daltons and both length and base composition may differ in two *Xenopus* females. There are about 450 ribosomal genes in each nucleolar organizer.

Since an amphibian oocyte contains more than 1000 nucleoli, it follows that the genes for nucleolar RNA synthesis must exist in many copies; that is, they are considerably *amplified.* When and where does the amplification of the genes which code for rRNA synthesis (or, more exactly, for the synthesis of the nucleolar precursor of the rRNAs) occur?

It has been proved that the amplification of the rRNA gene occurs in the *caps,* which are present at the pachytene stage of oogenesis and which have been depicted in Fig. 23. The DNA of the cap is replicated many times during this early stage of oogenesis; but it is not transcribed into RNA until a much later period, when vitellogenesis and rRNA synthesis begin. It is absolutely certain that the extra-chromosomal DNA which is present first in the caps, then in each nu-

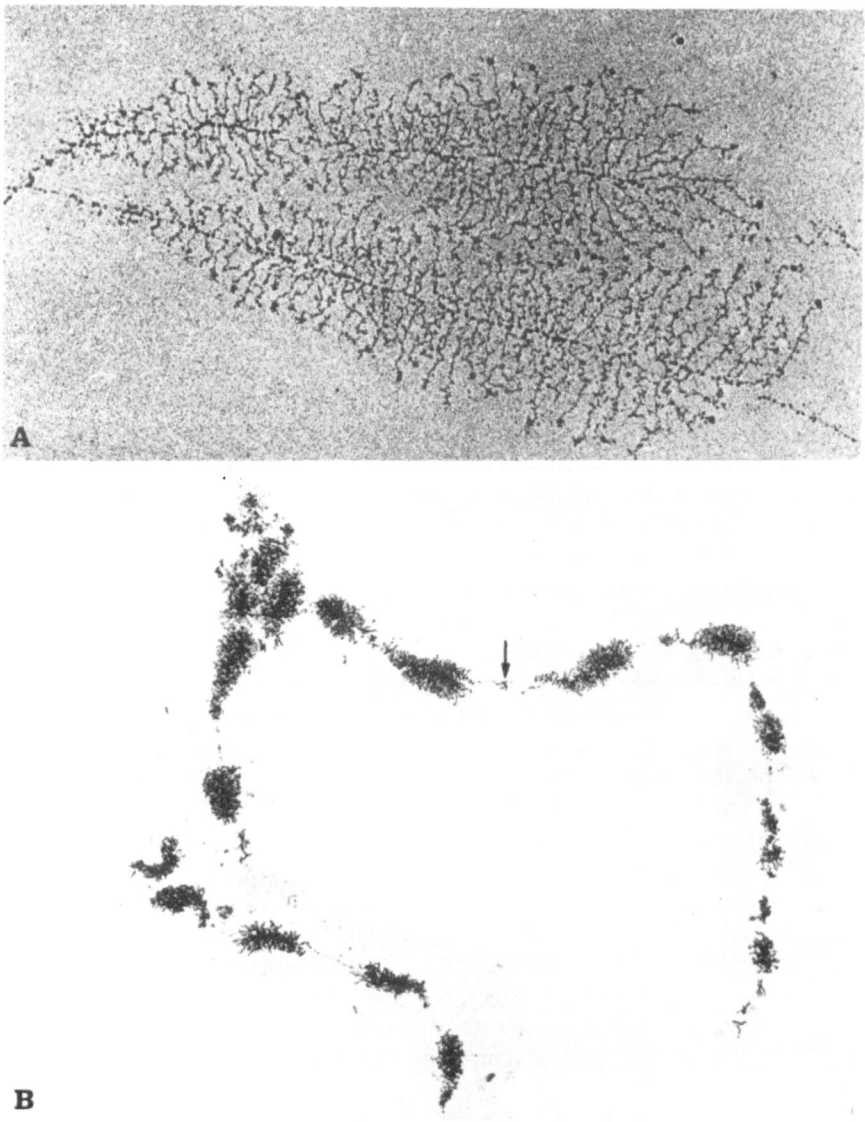

Fig. 27 A,B. Nucleolar organizers isolated from *Triturus* germinal vesicles and seen under the electron microscope. **A** Two transcription units ("Christmas trees"); **B** alternance of transcription units and spacers (*arrows*) at lower magnification. (Courtesy of Drs. O. L. Miller, Jr. and B. R. Beatty)

cleolus as the nucleolar organizer, is really localized in the caps; this has been clearly demonstrated by Pardue and Gall, who used *in situ* molecular hybridization. If a section, like that of Fig. 23, is first heated in order to denature the DNA (i.e., in order to separate the two strands of the double helix) and then treated with radioactive rRNA (28 S and 18 S), the latter hybridizes only with the DNA of the cap. This proves that the extra-chromosomal DNA of the cap and the rRNAs have many nucleotide sequences in common. In agreement with this conclusion, electron microscopy shows that the first nucleoli make their appearance in the cap. Each nucleolus contains some of the DNA which was present in the cap; this is the nucleolar organizer DNA present in each of the hundreds of nucleoli which are dispersed in the germinal vesicle.

Electron microscopy has disclosed the mechanisms of ribosomal gene amplification in the caps and has shown that unusual replication intermediates are present (DNA circles and "rolling circles" as in bacteria and phages). The replication mechanisms during ribosomal gene amplification are thus complex and very different from those used for chromosomal DNA replication prior to cell division (see Chap. VI). However, both are catalyzed by the same enzyme, DNA polymerase α. A vitellogenic *Xenopus* oocyte possesses as many as 2×10^6 copies of the ribosomal genes as compared to about 900 in an ordinary somatic cell (liver cell, red blood cell, etc.). The amplification of the ribosomal genes is thus at least 2,000 times, whereas simultaneously, the number of the nucleoli increases from 2 to about 1,200. Despite this huge amplification of the ribosomal genes, there is, as we have seen, no rRNA synthesis before vitellogenesis.

The nucleolar organizer DNA (which we can now call *ribosomal DNA* or rDNA) contains more $G + C$ sequences than chromosomal DNA. Due to this peculiarity and to gene amplification, it has been possible to isolate rDNA by ultracentrifugation as a heavy satellite which hybridizes specifically with radioactive 28 S and 18 S rRNA. This has allowed Brown and Dawid, as well as Birnstiel (1968), to study the molecular organization of the rRNA molecule (Fig. 28) in detail and to analyze the steps which lead from the nucleolar rRNA precursor (5×10^6 daltons) to the mature cytoplasmic rRNAs (1.3×10^6, 0.6×10^6 and 4.5×10^4 daltons). Only a small part of the spacer (600,000 daltons) is transcribed by RNA polymerase I, but this transcribed piece of the spacer is degraded during the processing of the rRNA precursor in the nucleus itself, as can be seen in Fig. 28.

In contrast to the 2000 times amplification of the 28 S and 18 S ribosomal genes, there is no amplification of the genes which code for the ribosomal proteins and for 5 S ribosomal RNA. We have already seen that there is no temporal coordination at the various stages of oogenesis between the synthesis of the 28 S, 18 S and 5.8 S RNAs on the one hand, and that of 5 S RNA on the other.

Although the 5 S genes are not located in the nucleolar organizers, but instead on several chromosomes, they deserve mention here. The first interesting point is that there are three different kinds of 5 S RNAs in *Xenopus:* in the *ovary,* a major and a minor 5 S RNA are encoded by respectively 24,000 and 2000 genes; *somatic* 5 S RNA differs from ovarian 5 S RNA by a few bases and it is coded by an-

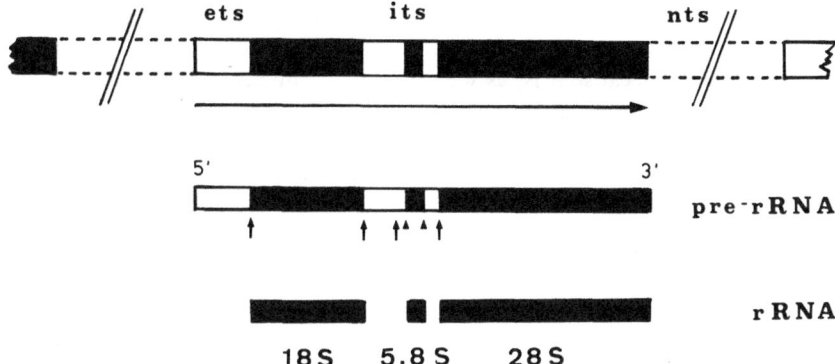

Fig. 28. Arrangement and transcription of rDNA (a ribosomal gene) into pre-rRNA (*large arrow*); processing of pre-rRNA into mature rRNA molecules. *ets* external transcribed spacer; *its* internal transcribed spacer; *nts* nontranscribed spacer. Primary (↑) and final (▲) cleavage of the pre-rRNA molecule

other set of 24,000 genes. All three 5 S RNAs have 120 bases (molecular weight 40,000 daltons); all of them are transcribed by RNA polymerase III. Another interesting point is the following: in previtellogenic oocytes, 5 S RNA is associated, in cytoplasmic ribonucleoprotein particles, to a 35,000 dalton protein which constitutes as much as 15% of the soluble proteins at this stage. This protein is of unusual interest, because it binds specifically to an internal region of the ovarian 5 S genes. This binding results in activation of these genes. Thus, binding to this protein stimulates the transcription of the ovarian 5 S genes by RNA polymerase III. This *transcription factor* disappears when the oocyte undergoes maturation and is absent in somatic cells, thus explaining why the ovarian 5 S genes are no longer active in fertilized eggs and somatic cells.

1.2.4 Nuclear Sap (Nucleoplasm)

After treatment with detergents and enzymes which remove the nucleic acids, oocyte nuclei retain, like somatic nuclei, a proteinaceous *nuclear matrix,* which forms a loose meshwork of fibers. In addition, the insoluble "skeletons" of the nuclear membrane and the nucleoli remain visible. The fibrillar meshwork is filled with a *nuclear sap* which contains many (at least 70) different proteins of cytoplasmic origin. Besides the karyophilic, nuclear specific, proteins, many others (actin, for instance) are common to both nucleus and cytoplasm. Actin is accumulated in the center of the germinal vesicle, where it forms a viscous "nuclear gel". It plays a role in maintaining the lampbrush configuration of the chromosomes, since injection of anti-actin antibodies is quickly followed by the condensation and inactivation of the lampbrush chromosomes.

A major protein of the germinal vesicle is *nucleoplasmin* which represents as much as 10% of all nuclear sap proteins. This nuclear-specific protein deserves special attention because it is, as we shall see, a "nucleosome assembly factor", which plays an important role in the assembly of DNA into chromatin.

The germinal vesicle contains 80–90% of total *DNA polymerase α* activity. This enzyme is thus accumulated in the nucleus despite the fact that there is no nuclear DNA synthesis once rDNA amplification is over. The three RNA polymerases (I, II and III) are also particularly abundant in the germinal vesicles of *Xenopus* oocytes which are widely used as a rich source of these RNA-synthesizing enzymes.

Finally, the nuclear sap contains small amounts of very heterogeneous RNAs. They are a complex mixture of primary transcription products, fragments resulting from the processing of RNA precursors, finished molecules of mRNAs, rRNAs and tRNAs on their way for export to the cytoplasm, and small nuclear RNAs (snRNAs).

1.2.5 Xenopus Oocytes as Test-Tubes

Thanks to J. Gurdon, *Xenopus* oocytes have become wonderful test tubes for molecular and cell biologists. Injection of any material into either the cytoplasm or the nucleus of these large (1.2 mm in diameter) cells does not require exceptional manual skill or costly equipment.

We have already seen that microinjection of labelled proteins into oocyte cytoplasm has provided very important data about the nuclear membrane "permeability". We have also seen that *Xenopus* oocytes are an exceedingly favorable material for the study of mRNA translation, which takes place after injection of any kind of mRNA into the cytoplasm. They lend themselves very well, in addition, to the analysis of the complex steps which lead to the excretion of secretory proteins (albumin, insulin, immunoglobulin, etc.). Recent experiments in which two different mRNAs were injected together into the cytoplasm of a *Xenopus* oocyte, have shown that one of the two may be preferentially translated. The continuation of this type of experiment should lead to a better understanding of the factors controlling the specificity of mRNA translation in the cytoplasm.

Pioneer experiments, in Gurdon's laboratory, have shown that injection of a *pure gene* into the germinal vesicles of *Xenopus* oocytes is followed by intense and prolonged (lasting several days) *transcription* of the gene. This is made possible by the already mentioned fact that the three major forms of RNA polymerases are accumulated in the oocyte nucleus. If, for instance, the chick ovalbumin gene is injected into the oocyte nucleus, the whole sequence is transcribed. The primary transcript is then processed in the oocyte nucleus, whereby seven introns are accurately removed and a (poly A) sequence is added to the 3′ end. The resulting ovalbumin mRNA moves into the oocyte cytoplasm, where it is translated, and finally chick ovalbumin is secreted in the medium by an amphibian oocyte. If one deletes, by in vitro chemical manipulations, the sequences flanking the 5' end of the gene, normal transcription still takes place. These control sequences, which are required for in vitro transcription are thus not needed for in vivo transcription in the oocyte nucleus. Curiously, if one injects into the nucleus, instead of the ovalbumin gene, the ovalbumin cDNA (a DNA copy obtained by treating ovalbumin mRNA with the viral enzyme reverse transcriptase), it is correctly transcribed, but there is no ovalbumin synthesis in the cytoplasm. The only difference between the cDNA and the gene is that the former has no introns. Thus, removal of the introns, by splicing, seems to be necessary to allow the passage of ovalbumin mRNA from the nucleus to the cytoplasm (Wickens et al. 1980). Similar approaches have led to great progress in our understanding of the control of sea urchin histone gene transcription and of the complex processing of tRNA precursors. Today we can induce at will, by in vitro chemical manipulations, specific mutations at a given site of a DNA molecule (in the gene or in its flanking sequences) and test their effects by injecting the modified DNA into the oocyte nucleus. Thus, *site-directed mutagenesis* opens possibilities which the classical induction of random mutations by mutagens could not give. A new kind of genetics is born which should lead to spectacular progress in our understanding of gene regulation in eukaryotes.

Injection of a DNA molecule into the germinal vesicle of a *Xenopus* oocyte may have several consequences. In experiments done by Trendelenburg and Gurdon (1978), ribosomal genes attached to a plasmid DNA were injected with the following results. The ribosomal genes were transcribed, as shown by their now familiar "Christmas tree" configuration; plasmid DNA was organized in nucleosomes, while spacer DNA was not. The germinal vesicle thus contains factors which allow the assembly of DNA molecules in nucleosomes (*chromatin assembly factors*). The main factors involved in nucleosome assembly are the enzyme DNA-topo-isomerase I and the karyophilic protein nucleoplasmin; both are abundant in the oocyte nucleus. It seems that the injected DNA molecules are first relaxed by topo-isomerase I, which would allow DNA to react with a histone store accumulated in the nucleus. The role of nucleoplasmin would be to render the nucleosomes insoluble in the nuclear sap environment.

Finally, Gurdon performed a series of very interesting experiments on the *controls exerted by the cytoplasm on nuclear activities.* He first injected *Xenopus* adult brain nuclei into the cytoplasm of *Xenopus* oocytes and eggs (1968) and obtained the following results: if brain nuclei are injected into full-grown oocytes, they swell and synthesize RNA. However, if they are injected into oocytes undergoing maturation, their chromatin condenses and they may even form chromosomes and mitotic spindles; they no longer synthesize RNA. Finally if brain nuclei are injected into unfertilized eggs, RNA synthesis stops, but DNA synthesis now takes place. Evidently, the injected adult nuclei behave exactly like the oocyte nucleus at these stages of development. Cytoplasmic factors decide whether the foreign nucleus will swell or condense and whether it will synthesize DNA, RNA or no nucleic acid at all. Recent improvements in the methods available for protein separation have allowed Gurdon and colleagues to go further. They injected nuclei from the human cancer HeLa cells into large *Xenopus* oocytes and analyzed the newly synthesized proteins. They found that only three HeLa proteins (of 25) were synthesized by the transplanted nuclei. Thus, the oocyte cytoplasm has the remarkable property of *reprogramming* the expression of individual HeLa genes. In other experiments, De Robertis and Gurdon (1977) injected nuclei from *Xenopus* adult kidney into oocytes of an urodele, *Pleurodeles*. Analysis of the proteins showed that the injected adult *Xenopus* nuclei synthesize some of the *Xenopus* oocyte typical proteins, whereas simultaneously, the synthesis of the specific kidney proteins stops. The cytoplasm of the oocyte thus contains factors which can reactivate genes, which were active during oogenesis, and inactivate genes, which were active in the adult. Finally, Korn and Gurdon (1981) have injected adult *Xenopus* nuclei into the germinal vesicles of *Xenopus* oocytes and followed 5 S RNA synthesis. Thus we have seen that the ovarian 5 S RNA genes are inactive in somatic cells, in which another set of genes directs the synthesis of 5 S RNA. The result was that ovarian 5 S RNA synthesis took place after injection into the germinal vesicle of some (but not all) oocytes. The nucleus of these oocytes contains a factor (probably a protein) which can reactivate the 5 S ovarian genes which were silent in the adult nuclei. There is no reactivation of these genes in oocytes which lack this protein factor.

2 Maturation of *Xenopus* Oocytes

Maturation is the progression of meiosis from the oocyte to the fertilizable egg stage, a stage which varies among different animal species. In amphibians, only the first polar body is expulsed (see Fig. 10). However, completion of meiosis does not take place until the egg has been fertilized or "activated" by pricking with a glass needle.

Morphologically, the most spectacular process is *germinal vesicle breakdown* (GVBD). In *Xenopus,* the disintegration of the nuclear membrane starts at the basal pole of the GV and spreads toward the animal pole [Fig. 19(3)]. Simultaneously, the lampbrush chromosomes strongly condense and attach themselves to a spindle; the nucleoli disintegrate, leaving only the nucleolar organizers. Both spindle and nucleolar organizers move toward the animal pole; the first meiotic division follows and the first polar body, which will degenerate, is expulsed (Fig. 29). The nucleolar organizers remain in the oocytes where they slowly disintegrate.

Maturation is *induced,* in many animal species, by *hormonal stimulation.* In amphibians, the natural inducer is *progesterone,* which is synthesized by the follicle cells under pituitary hormone stimulation. Maturation is easily obtained if oocytes, surrounded by their follicle cells, are dissected and treated in vitro with either progesterone or a pituitary extract. In the first case, addition of actinomycin D does not prevent GVBD, demonstrating that progesterone-induced maturation does not require RNA synthesis. On the other hand, pituitary hormone-induced maturation is actinomycin D-sensitive: production of progesterone by the follicle cells requires RNA synthesis. This is the rule when steroid hor-

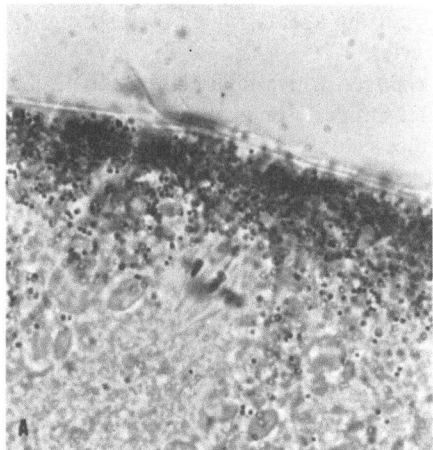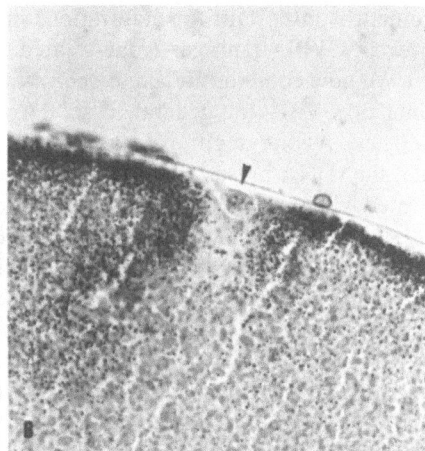

Fig. 29 A, B. Maturation of *Xenopus* oocytes. First (**A**) and second (**B**) maturation spindles. *Arrowhead:* first polar body

mones stimulate target cells (those of the uterus, for instance); thus, progesterone-induced maturation provides an interesting exception to the rule that response to steroid hormones requires RNA synthesis.

At variance also with the response of somatic target cells to progesterone is the fact that the hormone must act directly on the *cell surface*. It is inactive if injected into the oocyte, except if it has been dissolved in an oil droplet instead of water. Derivatives of steroid hormones which are unable to cross the cell membrane are capable of inducing maturation.

It came as a surprise when we found, in 1975, that organomercurials, which block –SH groups without penetrating into the cells, induce GVBD. Since that time, many other substances, totally unrelated to steroids, have been found to induce maturation as efficiently as progesterone (lanthanum chloride, propanolol, insulin, etc.). All these agents share one property: they set membrane-bound calcium free and increase the free Ca^{2+} concentration in the cytoplasm. In fact, increasing the free Ca^{2+} content by addition or slow injection of $CaCl_2$ is enough to induce maturation. Recently, it has been shown that there is indeed a fast Ca^{2+} burst when progesterone is added to *Xenopus* oocytes. However, it would be unwise to believe that Ca^{2+} is the only ion involved in maturation: magnesium, manganese and potassium ions can, under certain circumstances, induce maturation in *Xenopus* oocytes.

The fast increase in the free Ca^{2+} content of the oocyte is not the only early chemical response of the oocyte to progesterone. Another fast change, which precedes GVBD by several hours, is a strong inhibition of membrane *adenylate cyclase,* the enzyme which synthesizes cAMP at the expense of ATP. As a result, the cAMP content drops sharply between 15 and 60 s after progesterone addition and cAMP comes back to its initial level within 1 h. This initial drop in the cAMP content is important for maturation, since all the agents, which increase cAMP prevent GVBD in progesterone-treated oocytes.

We now come to the *late biochemical changes,* which take place at about the time of GVBD (in general, 6 to 8 h after progesterone addition to *Xenopus* oocytes). As one might expect, rRNA synthesis stops when the nucleoli degenerate: the rDNA present in the nucleolar organizers is neither transcribed nor replicated when they are casted in the cytoplasm. There have been reports that synthesis of the other kinds of RNAs continues during maturation. If so, these RNAs are not required for maturation itself, since we already know that this process is actinomycin D-insensitive. However, they might play a role in processes, which follow fertilization, but this is still a hypothesis.

Progesterone treatment induces a progressive increase in *oxygen consumption* and *protein synthesis* in *Xenopus* oocytes; both reach a peak at the time of GVBD. The marked increase in protein synthesis is more of a quantitative than a qualitative nature: among 600 proteins, only 5 selectively increase during maturation. However, there is an interesting selectivity in the stimulation of histone synthesis: "core" histones (i.e. H_2A, H_2B, $H3$ and $H4$ histones) increase 20 to 50 times, while the synthesis of $H1$ histone remains unaffected by progesterone treatment. A very important point is that this increase in protein synthesis, including the selective increase in core histones synthesis, takes place also in oocytes which have been *enucleated* before progesterone administration. This shows that protein synthesis, at maturation, is controlled by *selective post-transcriptional mechanisms* which remain poorly understood.

Studies with *inhibitors* of energy production and protein synthesis have disclosed an interesting fact. They prevent GVBD if they are allowed to act during the first 3 h after progesterone addition, but they have no effect after that time. Protein synthesis and energy production are thus absolutely required during an early period of maturation; but GVBD itself does not require them any more.

There is no *DNA synthesis* during maturation (except a low level of mitochondrial DNA synthesis). However, maturation is the stage of development, in which the *machinery* necessary for DNA synthesis is built up: the overall activity of DNA polymerase α increases four times during maturation and several enzymes and factors involved in DNA synthesis make their appearance. This building up of the DNA-synthesizing machinery, as well as the increase in core histones, are a preparation for egg cleavage, in which the number of nuclei greatly and quickly increases. Curiously, chromosomal DNA and rDNA in maturing oocytes are efficiently protected against this powerful DNA-synthesizing machinery, in contrast to foreign DNA molecules injected into maturing oocytes.

Shortly before GVBD, there is a strong wave of *protein phosphorylation:* many proteins, both soluble and membrane-bound, are simultaneously phosphorylated. One of the targets for protein kinases might be the pore-lamina complex proteins of the nuclear membrane. Their phosphorylation might conceivably lead to the disintegration of this membrane at GVBD. However, the wave of protein phosphorylation occurs even in oocytes which have been enucleated before progesterone treatment.

The changes in cAMP content and in protein phosphorylation which take place soon after progesterone stimulation have led to a *hypothetical model* of the molecular events which underlie maturation. The initial Ca^{2+} liberation and the drop in cAMP content would lead to the dephosphorylation of a key phosphoprotein present in full-grown oocytes; maturation would be prevented so long as this protein is in its phosphorylated form. Dephosphorylation of this protein probably results from the activation of a protein phosphatase present in the oocytes. The burst of protein phosphorylation which shortly precedes GVBD and which is believed to be responsible for the rupture of the nuclear membrane would result from an increased activity of a cAMP, or (more likely), a Ca^{2+} dependent protein kinase. This increase in protein kinase activity would also be responsible for chromosome condensation, the target being this time histone H 1.

Analysis of maturation in *Xenopus* oocytes has led to the discovery of several biological "factors" of still unknown chemical mature. The most important is the *"maturation promoting factor"* (MPF): injection of cytoplasm, taken from a progesterone-treated oocyte a few hours before GVBD, into a normal recipient oocyte induces GVBD within 1–2 h (thus, much faster than after progesterone stimulation). The factor responsible for the induction of GVBD (breakdown of the nuclear membrane, condensation of the chromosomes, disappearance of the nucleoli) has been called *MPF* (maturation promoting factor); its injection into full-grown oocytes induces a rapid increase in protein synthesis and in protein phosphorylation. When MPF-containing cytoplasm is injected into a recipient oocyte, this oocyte synthesizes more MPF by a kind of autocatalytic amplification. MPF synthesis occurs in enucleated progesterone-treated oocytes; its formation requires protein synthesis during the first hours which follow addition of the hormone. MPF has no species-specificity: injection of MPF-containing cytoplasm from *Xenopus* oocytes into starfish oocytes induces GVBD. This factor disappears from the cytoplasm at the end of maturation, but it reappears in a cyclic way, during egg cleavage at the time in which nuclear membrane breakdown and chromosome condensation occur. It is probable that there is a reciprocal cycling, during cleavage, of MPF and an agent which inactivates it. MPF-like factors have recently been found in HeLa and other somatic cells. Their role is to induce, as in oocytes, nuclear membrane rupture and chromosome condensation. Unfortunately, we know next to nothing about the chemical nature of MPF and its mode of action: all we know is that it is a protein which probably involved in protein phosphorylation.

At the end of maturation, when the first polar body has been expulsed, a second factor makes its appearance in the oocyte cytoplasm. It is a *cytostatic factor* (CSF): if one injects cytoplasm from an unfertilized egg into one of the blastomeres of a cleaving egg, mitotic activity is arrested at metaphase, in the injected blastomere. This arrest of mitotic activity is probably due to an inhibition, by CSF, of the depolymerization of the spindle microtubules at anaphase (see Chap. VI). In the unfertilized egg, the physiological role of CSF is probably to arrest meiosis at metaphase of the second meiotic division.

Does the oocyte nucleus (the GV) play any role during maturation? We have seen that progesterone induces, in enucleated oocytes, as well as in normal ones, an increase in protein synthesis (including the preferential synthesis of the core histones) and a burst in protein phosphorylation; both produce, at the same time, the all-important MPF. The same changes in ionic permeability the same morphological changes (retraction of the microvilli, opening up of the cortical granules after pricking with a needle) occur in progesterone-treated nucleate and enucleated oocytes. However, experiments in which nuclei from adult cells and spermatozoa, or centrioles were injected into nucleate and enucleated progesterone-treated oocytes have shown that the mixing of the nuclear sap with the surrounding cytoplasm has important consequences. In enucleated oocytes, in contrast to nucleate ones, the chromatin of injected somatic nuclei is unable to condense and the nuclei of spermatozoa do not swell. In such oocytes, in contrast with nucleate oocytes, injection of centrioles is never followed by the formation of asters around them. These experiments lead to the conclusion that mixing of the nuclear sap with the cytoplasm is necessary to induce a *cytoplasmic maturation* more discrete than the morphologically visible nuclear maturation.

We have been recently interested in the induction of maturation in small, still *vitellogenic,* oocytes (0.6–0.8 mm in diameter). They do not respond to progesterone, but GVBD can be obtained by injecting, into such young oocytes, MPF-containing cytoplasm taken from large progesterone-treated oocytes. However, maturation is abortive: young oocytes, after MPF injection, break down their nuclear membrane and condense their chromosomes, but they are unable to build up a meiotic spindle. Thus, the chromosomes remain scattered in the cytoplasm. Experiments in which the contents of such MPF-injected young oocytes have been injected into full-grown, progesterone-treated oocytes have shown that the chromosomes of the young oocytes are not protected, as those of the full-grown oocytes, against the DNA-synthesizing machinery which is built up during maturation. The chromosomes of the young oocytes multiply and attach themselves to a spindle (which they are unable to build when they are in their own young cytoplasm).

Finally, we should add that what has been said in this section about *Xenopus* oocytes remains essentially valid for other animal species, even when the initial maturation-inducing stimulus is provided by other agents than progesterone.

In eggs of many species (mammals, fishes, starfishes) a role for calcium ions, cAMP content, protein synthesis and protein phosphorylation has been observed and the appearance of MPF seems to be a very general, if not universal, phenomenon. Maturation, which is a very important stage of development, thus follows the same general course in the whole animal kingdom. There is, of course, some *diversity* in both morphology and biochemistry when one compares different species, but there is also a remarkable *unity* in the achievement of the final result, the production of a fertilizable egg.

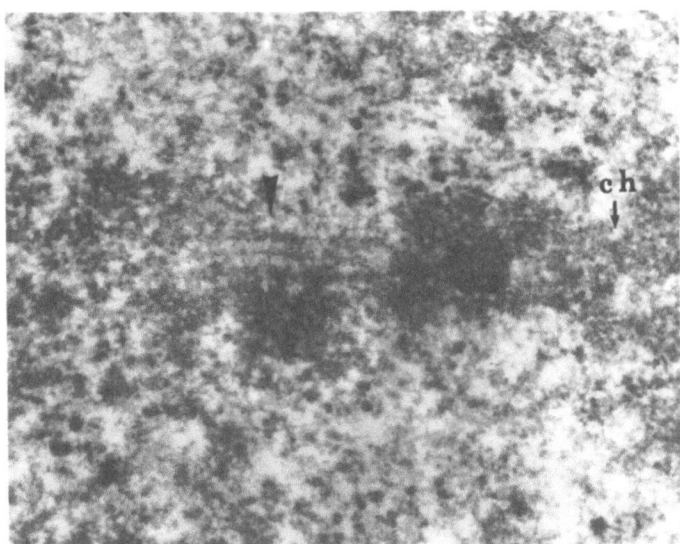

Fig. 30. Synaptonemal complex in a *Xenopus* oocyte at the pachytene stage (*arrowhead*). ch chromosome. (Courtesy of Prof. P. Van Gansen)

3 Spermatogenesis

The reason why a cell undergoes meiosis rather than mitosis remains completely unknown despite the fact that spermatogenesis often provides more favorable material for the study of meiosis than oogenesis. One of the factors which play a role in holding the homologous chromosomes together when they undergo pairing is the existence of *synaptonemal complexes* (Fig. 30). These curious structures are found only in the *chiasmata* which link the homologous regions of the paired chromosomes at the time of crossing-over. Inhibitors of protein synthesis produce the disappearance of the synaptonemal complexes and the paired chromosomes separate from each other. This shows that these complexes, which are made of proteins and RNA, but apparently do not contain DNA, play a role in keeping the paired chromosomes together. However, it is not likely that the synaptonemal complexes, which are too long, for that purpose have the specificity required for controlling the very precise DNA exchanges which occur, at the molecular level, during crossing-over and which lead to genetic recombination. So far, only hypotheses devoid of serious experimental bases have been presented to provide explanations for gene recombination at meiosis. However, a few facts have been established. Since they are valid for both plants (lily anthers) and animals (mouse spermatogenesis), it is likely that they have a general significance. For instance, a special kind of DNA is synthesized after the last chromosomal DNA replication in spermatogenesis. This late replicating DNA, which is quantitatively of minor importance and is linked to proteins and lipids, might

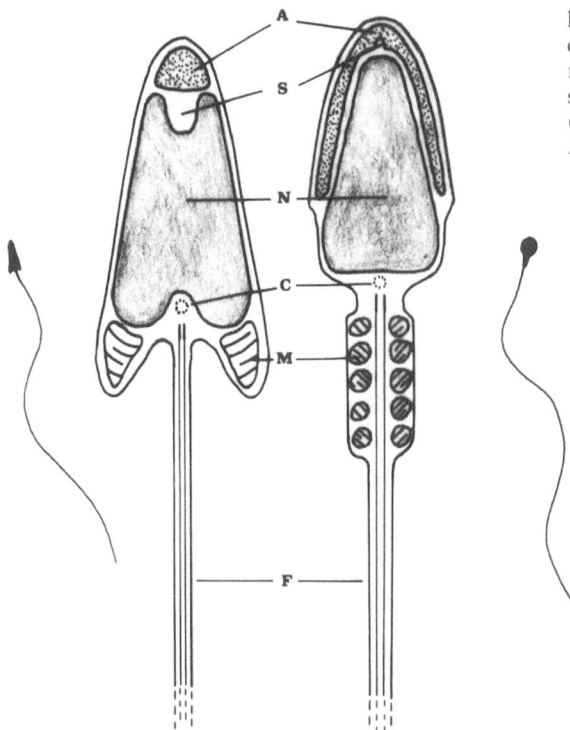

Fig. 31. A mature spermatozoon of an echinoderm (*left*) and a mammal (*right*). *A* acrosome; *S* subacrosomal space; *N* nucleus; *C* centriole; *M* mitochondria; *F* flagellum

play a role in the pairing of the meiotic chromosomes and in genetic recombination. Crossing-over and genetic recombination require DNA strand scission and repair. The evidence showing that the activity of the DNA repair enzymes reaches a peak at the time in which gene exchanges take place (pachytene stage of meiosis) is steadily increasing.

After the two meiotic divisions, the four haploid spermatids are transformed into mature spermatozoa (Fig. 10) by a process called *spermiogenesis.* During that period, most of the cytoplasm degenerates; the cytoplasmic RNA, in particular, is rejected by the spermatid and is finally degraded. The nucleus progressively undergoes condensation during spermiogenesis.

Figure 31 depicts the adult, mature spermatozoon. At the anterior extremity is a special granule, the *acrosome,* which originates from the Golgi bodies of the spermatid. It contains a number of acid hydrolases: acid phosphatase, acrosin (a proteolytic enzyme), hyaluronidase (which attacks the mucoproteins) etc. The acrosome, in view of its enzyme content, can be regarded as a *specialized lysosome.*

Behind the acrosome is *the subacrosomal space* (which plays a role in fertilization that will be described in the next chapter) and the *nucleus,* which forms the main part of the sperm "head". It contains the haploid (*n*) complement of DNA, corresponding to all the paternal genes. This paternal DNA is very strongly bound to exceedingly basic proteins. During spermiogenesis, the usual histones are progressively replaced by a special kind of "very arginine-rich" histone or even by protamines (polypeptides with a very high content in arginine). The result is that the DNA-basic protein complex becomes almost crystalline in the sperm nucleus. The DNA is now in a completely repressed, inactive form. There are no nucleoli in the sperm nucleus.

Behind the nucleus, one finds the *mid-piece,* which contains the *mitochondria.* They are always well developed, often being very numerous; when present in small number (one or two only), they are of very large size. They contain all the respiratory enzymes and are extremely active in oxidative phosphorylations. They actively oxidize either endogenous (phospholipids) or exogenous (fructose) substrates. Two *centrioles,* a proximal and distal one, are also present in the mid-piece; they have the same complicated structure under the electron microscope, but play different roles. The proximal centriole is probably necessary, as we shall see, for fertilization; the distal centriole lies at the basis of the flagellum. It is a basal body (see Fig. 8) and its role is to give rise to the flagellum. However, it is not required for flagellar motion which can take place in isolated cilia and flagella. The flagellum, which is usually surrounded by a spiral sheath of protein, forms the *tail* of the spermatozoon. Like all other flagella, it is, as a rule, made of nine outer and two central filaments. With the sperm flagella, we are again faced with the unity and diversity of living beings: all flagella have the same general structure, function (beating) and overall chemical composition (tubulin is their major protein). However, marked, species differences in the details of the ultrastructure have been described. In some species, there is a single central filament, in the eel, spermatozoa even completely lack the central filaments and, in mammals, flagella display peri-axonal structures, including nine dense fibers and two longitudinal columns.

The nine outer and two central filaments are, in fact, hollow microtubules, resulting from the linear polymerization of tubulin molecules. Motility is due to the sliding of the outer microtubules. This sliding requires energy, resulting from ATP hydrolysis by an adenosine triphosphatase (ATPase). Sperm flagella, as well as cilia, indeed possess a specific ATPase called *dynein;* it is located in the two "arms" attached to the outer microtubule doublets (see Fig. 8). In man, there are individuals, in which as the result of a genetic defect, the dynein arms are missing. These individuals are sterile, despite the fact that spermatogenesis is apparently normal: they produce spermatozoa, but since these spermatozoa lack dynein ATPase, they are immobile.

As one can see, the spermatozoon is a highly specialized cell. Having lost its ribosomes, it is unable to synthesize proteins; its condensed nucleus is in a completely repressed form, so that RNA synthesis is also absent in spermatozoa. The flagellum gives the necessary motility to the spermatozoon; the energy needed for movement comes from the very active mitochondria. The acrosome, as we shall see, helps the spermatozoon to reach the egg surface. One can only be struck by the analogy existing between a spermatozoon and a bacteriophage (Fig. 9). In both cases, we are dealing with machines that are highly specialized for the injection of DNA, the genetic material, into the recipient cell.

CHAPTER V

Fertilization: How the Sleeping Egg Awakes

1 General Outlook

Fertilization of the egg is an all-important event for the life of all organisms that reproduce sexually. Not only does it start the machinery for embryonic development, but it also has fundamental genetic consequences: the diploid ($2n$) state is restored and the paternal hereditary characters are introduced into the fertilized egg (also called the zygote). In most species, the sex of the future adult is established at fertilization.

Figure 32 summarizes the *morphological changes* that occur at fertilization. The egg, according to species, may be at any stage of maturation (first-order

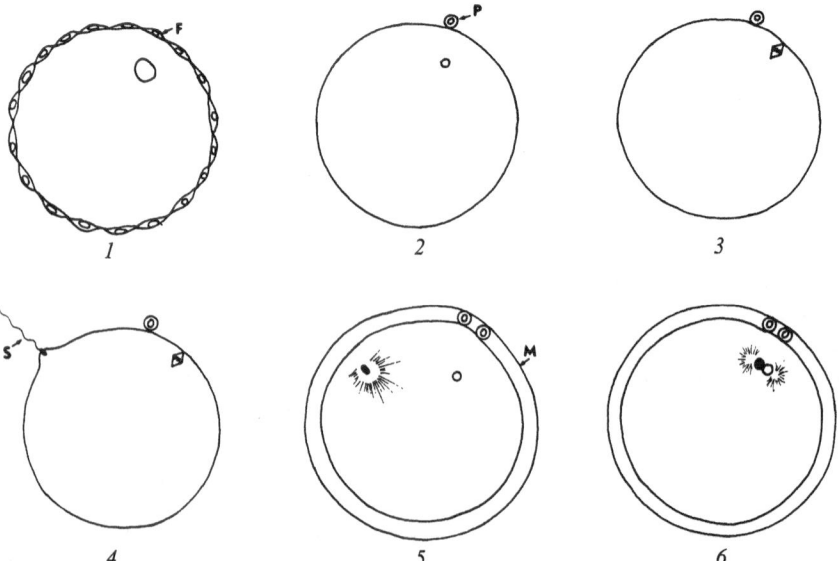

Fig. 32. Fertilization of an amphibian egg. *1* Unfertilized egg surrounded by the follicle cells (*F*). *2* Oocyte undergoing maturation after hormonal stimulation; the first polar body has been expelled (*P*). *3* Maturation is blocked at metaphase of the second meiotic division. *4* The ripe oocyte is fertilized by a spermatozoon (*S*). *5* Maturation is completed; the second polar body has been expelled and a fertilization membrane (*M*) surrounds the egg; the spermaster is forming around the male pronucleus. *6* The two pronuclei are fusing together (amphimixy)

oocyte with an intact germinal vesicle, metaphase of the first meiotic division, second-order oocyte which has expelled the first polar body; metaphase of the second meiotic division; ootide which has eliminated the two polar bodies). If maturation is not yet complete (or has not even started), the contact of the sperm head with the ovular cell membrane induces or completes maturation. As a first visible response, a *fertilization membrane* rises up and surrounds the egg; this membrane appears first at the point of entrance of the spermatozoon and progresses toward the opposite pole of the egg. Membrane elevation is fast (a few seconds) in small eggs, such as those of the sea urchin and much slower in large ones (1 h or more in frog eggs). Between the membrane and the surface of the egg is the *perivitelline space,* filled with the *perivitelline fluid.* The fact that the egg now bathes in a liquid medium allows it to undergo rotation under the influence of gravity. The animal pole now points upward, the heavy vegetal pole downward. At about the same time, astral rays begin to surround the sperm head, forming the *spermaster;* this aster then regresses and the sperm nucleus, which has now swollen and is called the *male pronucleus,* moves toward the egg nucleus (*female pronucleus*). The two pronuclei finally fuse together (*amphimixy*) and fertilization is complete. Two asters appear at the opposite ends of the *zygote* nucleus and a mitosis, which marks the beginning of the next stage, cleavage, immediately follows.

This is a very broad description of fertilization which is by no means a constant process and shows marked differences in the various animal species. We shall now go into a few details, taking as an example the *sea urchin* egg, which has been most extensively studied.

The first phenomenon which deserves mention is the *acrosome reaction* (Fig. 33 a, b). The acrosome of the spermatozoon opens when it comes in contact with the jelly coat that surrounds the egg and the inner acrosome membrane changes into a straight flexible filament which penetrates into the surface of the egg. This acrosome filament results from the linear polymerization of actin molecules, previously accumulated in the subacrosomal space (Fig. 31), as soon as the outer acrosome membrane has broken down. Simultaneously, acid hydrolases present in the acrosome are released in the medium and contribute to the digestion of the envelopes which protect the unfertilized egg (jelly coat, vitelline membrane). The acrosome reaction is unspecific: it is not induced by the jelly coat alone, but also by alkaline seawater, contact with glass, etc. Calcium fluxes and changes in the acrosome pH play an essential role in the acrosome reaction. However, binding of the spermatozoon to the egg surface results from a specific interaction between receptors present on the vitelline membrane which surrounds the unfertilized egg and a species-specific glycoprotein, called *bindin,* which is located on the acrosome membrane. This specificity explains why crosshybridization between different sea urchin species does not occur in nature. The penetration of the sperm nucleus into the egg is not due to a retraction of the acrosome filament, but to the fusion of the membranes of the acrosome vesicle and of the unfertilized egg (Fig. 33 c, d). Once these two membranes have fused together, fertilization can be regarded as successful and the sperm nucleus will be

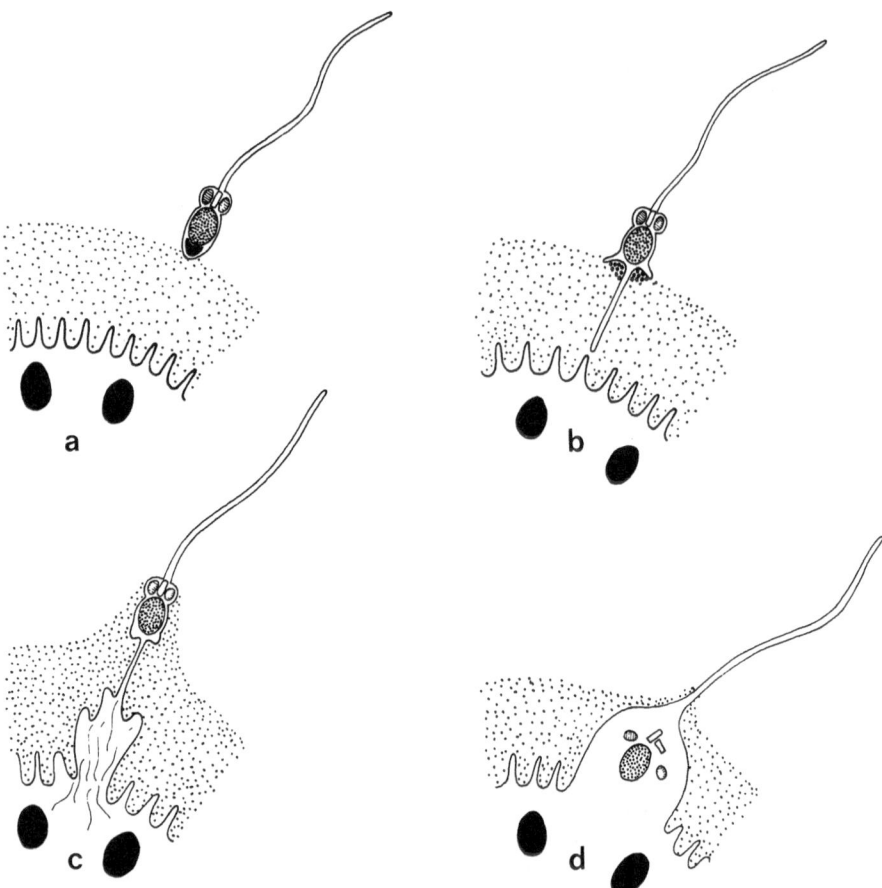

Fig. 33a–d. Fertilization as seen under the electron microscope. **a** A spermatozoon comes in contact with the jelly coat (*stippled*) which surrounds the egg. **b** The acrosome of the spermatozoon opens (acrosome reaction). **c** Fusion between the membranes of the oocyte and the acrosome. **d** The sperm nucleus is injected into the egg

drawn into the egg through the channel left between the fused membranes; cortical microfilaments of actin are involved in this process. The sperm nucleus still has its condensed chromatin and it is not surrounded by a nuclear membrane. But, after a few minutes, sperm chromatin begins to undergo decondensation; a nuclear membrane soon surrounds it, which is formed by the endoplasmic reticulum (inner membrane system) of the egg.

The lifting of the fertilization membrane, also called *cortical reaction* or *egg activation,* is accompanied by profound changes in the ultrastructure of the egg cortex. The microvilli, which were present in the oocyte, retract and the egg membrane becomes almost perfectly smooth. Simultaneously, there is a general

contraction of the egg. The cortical granules open up and their content is eliminated into the outer medium (seawater in the case of sea urchin eggs). The cortical granules contain glycoproteins of strongly acidic character (acid polysaccharides or mucopolysaccharides, containing very acid SO_3H groups). These
glycoproteins precipitate when they come in contact with seawater and immediately undergo polymerization. The result is the formation of the almost crystalline fertilization membrane. Just outside the egg membrane, a thin extracellular mucous layer expands all around the surface of the egg. This is the *hyaline
layer,* which is believed to act as a barrier against polyspermy, i.e. the penetration
of more than one spermatozoon in the egg. Polyspermy is a physiological
phenomenon in large eggs (for instance, those of urodeles) where the cortical reaction is slow. However, even here, only one sperm nucleus fuses with the egg
nucleus and all the other sperm nuclei finally degenerate. Various agents inhibit
the formation of the fertilization membrane. Especially useful for the experimenter, who wishes to work with "naked", membraneless eggs, are trypsin (a
proteolytic enzyme) and dithiothreitol (an –SH containing reagent which breaks
the –S–S– bonds present in proteins). In order to induce the cortical reaction, the
spermatozoon must come in contact with the egg membrane. If one injects a
spermatozoon into a sea urchin egg, both the egg and the sperm remain inert. If
one adds sperm to such a spermatozoon-injected egg, a normal fertilization
membrane forms. An egg that has been artificially fertilized by injection of sperm
can thus be refertilized, and the egg will be dispermic. Refertilization can also be
obtained by dissolving the hyaline coat of a fertilized egg by treatment with al-
kali or with calcium-free seawater, and then adding sperm again.

The *spermaster* is a powerful monaster which forms around the proximal centriole of the spermatozoon. There are no microtubules in unfertilized sea urchin
eggs, but they contain a large store of unpolymerized tubulin molecules. The
proximal centriole acts as a seed for tubulin polymerization and formation of astral microtubules. The role of the spermaster is to favor the migration of the
sperm nucleus (male pronucleus) toward the center of the egg where the egg nucleus (female pronucleus) lies. Inhibition of microtubule polymerization by treatment with drugs like colchicine prevents the migration and the fusion (amphimixy) of the two pronuclei.

The frequent occurrence of cytasters in eggs can throw some light on the formation of the spermaster. Asters never form in the cytoplasm of immature
oocytes, but small asters can often be seen when the nuclear membrane breaks
down during maturation, when nuclear sap and cytoplasm mix together. The appearance of hundreds of asters in the cytoplasm of unfertilized or just fertilized
eggs can easily be induced experimentally by treating the eggs with heavy water
(deuterium oxide: D_2O) (Fig. 34). These cytasters appear simultaneously and do
not divide afterwards. In the center of each aster, a small granule, which is probably a very tiny centriole (precentriole) can be seen with the electron microscope.
Since the cytasters can appear after D_2O treatment under conditions in which
protein, RNA and DNA synthesis are suppressed, it is likely that they result from
the self-assembly of pre-existing protein subunits previously dispersed in the

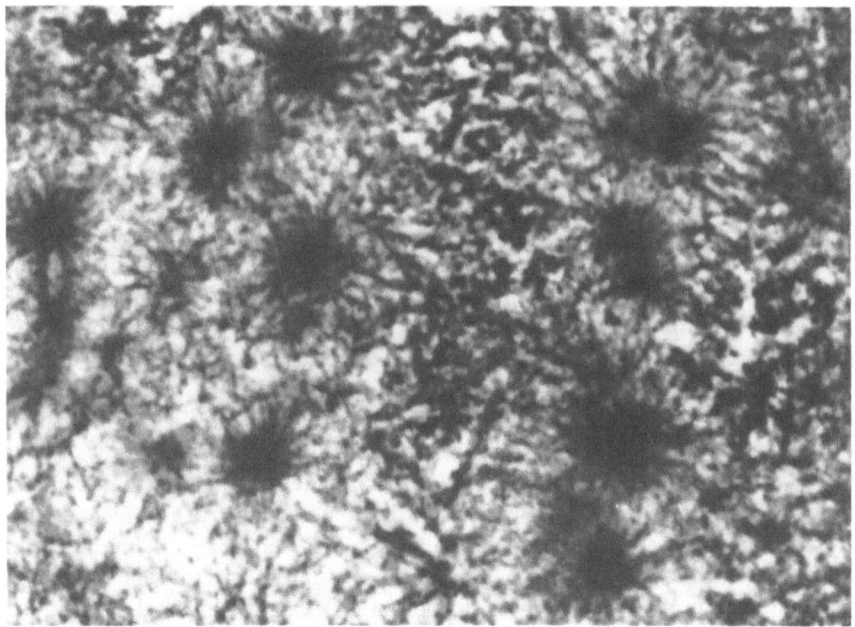

Fig. 34. Appearance of cytasters in an unfertilized amphibian egg treated with heavy water (D_2O) (Van Assel and Brachet 1966)

cytoplasm. Unfertilized eggs, in contrast to spermatozoa, lack visible centrioles, but the appearance of a large number of cytasters after treatment with agents which, like D_2O, stabilize the microtubules, shows that eggs possess the capacity to build up asters around materials which pre-existed in a diffuse form in their cytoplasm. As we shall see, authentic centrioles, surrounded by large asters, form when unfertilized eggs are treated with parthenogenetic agents which induce repeated cleavage cycles and later, full embryonic development (for instance, successive treatments with butyric acid and hypertonic seawater).

We know very little, except that microtubule activity is required, about the factors that control amphimixy, the fusion of the two pronuclei. Past experiments by A. Brachet, on artificial polyspermy in frog eggs have shown that the attraction between the two pronuclei does not depend on their "sex". If many spermatozoa (100 or more) penetrate into an egg that has been subjected to treatments which slow down the cortical reaction, adjacent sperm heads can fuse by groups of two or three. Such strongly polyspermic eggs never undergo cleavage afterwards. It is likely that, in normal fertilization, the spermaster helps the migration of the male pronucleus toward the female one.

Regarding the *swelling of the male pronucleus,* it can be said that it is controlled by the conditions prevailing in the cytoplasm. A. Brachet succeeded, long ago, in fertilizing sea urchin eggs that had not yet completely finished their maturation; he found that if the egg chromosomes were in a condensed state, the male pronucleus remained condensed. If, on the other hand, the egg nucleus had swollen, the spermatozoon nucleus quickly underwent swelling. Similar results have been obtained, more recently, by Gurdon who, as we have seen, injected nuclei into amphibian oocytes at various stages of maturation. It seems clear today that nuclear swelling is not simply the result of hydration of the nucleus, but that it is due to the uptake of cytoplasmic proteins into the nucleus and to the loss of the sperm-specific basic proteins.

We can now turn to the physical and chemical changes that occur at fertilization; in the following, we shall be dealing with sea urchin eggs unless otherwise stated.

2 The Fertilizin Problem

Early in this century, Lillie observed that seawater in which unfertilized sea urchin eggs have been left for some time ("egg water") exerts various effects on the spermatozoa. It agglutinates the spermatozoa by their heads, increases their motility and keeps them active for a longer time. The egg water thus contains a factor (*fertilizin*) which increases the activity of the spermatozoa. For Lillie, fertilizin would even attract the spermatozoa and would thus have a chemotactic activity, but this was soon shown to be incorrect. Later work has demonstrated that spermatozoa release into the surrounding seawater substances that have biological effects.

Among these substances the most important are Lillie's *egg fertilizin* and *sperm lysin*. Fertilizin is simply the jelly coat which surrounds unfertilized sea urchin eggs. Many of its effects (induction of the acrosome reaction, increase in sperm motility and respiration) are unspecific and due to binding of the acid mucopolysaccharides constitutive of the egg jelly to heavy metals present in seawater which have adverse effects on spermatozoa. The only specific effect of fertilizin is the agglutination of spermatozoa, a phenomenon which contributes to the prevention of polyspermy if the eggs happen to be in the presence of a very large excess of spermatozoa.

Extracts of spermatozoa dissolve the jelly which surrounds and protects the eggs; this has been ascribed to a *sperm lysin* which is probably only the hydrolytic enzymes present in the acrosomes (hyaluronidase, which dissolves mucus, in particular).

3 Physical and Chemical Changes at Fertilization

In the following, we shall be primarily interested in sea urchin eggs, whereby one should draw a distinction between early and late changes at fertilization, as pointed out by D. Epel in 1978.

3.1 Early Changes

The first recorded change at fertilization is a brief *depolarization of the egg membrane potential;* it is believed to be due to an influx of Na$^+$ ions and to be responsible for the so-called fast block to polyspermy. This fast electrical block prevents the entry into the egg of more than one spermatozoon in *Xenopus* as well as in sea urchin eggs. A slower block to polyspermy is correlated with the elevation of the fertilization membrane which prevents the entry into the egg of supernumerary spermatozoa.

Another important early ionic change is an *increase in free Ca^{2+}*, much larger than that which follows progesterone addition in *Xenopus* oocytes (see preceding chapter). This increase is responsible for the breakdown of the cortical granules and the exocytosis of their contents. Increasing artificially the free Ca^{2+} content of unfertilized eggs, by treatment with a *Ca^{2+} ionophore* called A23187, is an excellent method for inducing a cortical reaction in eggs of many animal species. Activation of sea urchin eggs (chromosome condensation, DNA replication, formation of asters) can also be obtained by treatment of sea urchin eggs with *ammonia.* Ammonia bypasses the calcium rise step: there is no cortical reaction. The cortical granules remain intact and a fertilization membrane does not develop around the egg.

Ammonia penetrates easily into unfertilized sea urchin eggs and increases their *internal pH.* Normal fertilization also induces, within 1 and 4 min after sperm addition, an increase in the sea urchin egg internal pH as high as 0.3–0.5 pH units. This rise is due to an exchange between extracellular Na$^+$ and intracellular protons (H$^+$) which are ejected from the egg into the outer medium. A similar increase in internal pH at fertilization has been found in *Xenopus* eggs. In both cases, the internal pH remains constant during cleavage after its initial rise shortly after fertilization.

As we shall see in more detail, the classical method for inducing *parthenogenesis* in sea urchin eggs is that of Loeb. This involves two successive steps: treatment of the unfertilized eggs first with hypertonic seawater and then with butyric acid, or vice versa. With this two-step method, asters form around typical centrioles. With the other activation methods (treatment with the A23187 ionophore or with ammonia) asters also form, but the microtubules originate from the chromosomes and from amorphous osmiophilic foci. It seems that self-assembly of typical centrioles is a prerequisite for full parthenogenetic embryogenesis. In contrast, eggs activated by A23187 or ammonia treatment, stop developing after a few abortive cleavages.

3.2 Later Biochemical Changes

3.2.1 Oxygen Consumption

Warburg discovered, more than 70 years ago, that the respiration of unfertilized sea urchin eggs increases considerably (three to four times) a few minutes after the addition of sperm. However, this is not a general phenomenon, present in all eggs. If one compares the oxygen uptake of the fertilized and unfertilized eggs of a number of animals, one finds that respiration can either increase (as in the sea urchin), remain constant or even decrease at fertilization. Fertilized eggs of all the species studied (which have about the same size and all live in seawater) have almost the same oxygen consumption; on the other hand, the respiration of the unfertilized eggs of the various species show very great variations. The conclusion that can be drawn from these experiments is the following: fertilization does not necessarily increase the respiration of the egg, but it *regulates* this process. The respiration of the unfertilized egg, which is an inactive cell, is often abnormal; fertilization brings it back to the normal value.

That this conclusion is correct for sea urchin eggs has been shown by delicate experiments which made it possible to measure the respiration of a small number of carefully selected, perfectly synchronous oocytes. The oxygen consumption of the oocyte (which still has an intact germinal vesicle) is high; it falls considerably in the unfertilized egg (which has eliminated its two polar bodies) and remains low for a considerable time; it is restored to the initial high level by fertilization. What is peculiar to the sea urchin egg is not so much the strong increase in respiration, which follows fertilization, but the very low respiratory rate of the unfertilized egg.

Despite many efforts, the identity of the substrate which is oxidized by sea urchin eggs at fertilization has long remained mysterious. It is only recently that it has been discovered that this substrate is simply – water. The large and sudden increase of respiration at fertilization is mainly due to the formation of hydrogen peroxide (H_2O_2) which is used by an egg *ovoperoxidase* for hardening the fertilization membrane; H_2O_2 also inactivates spermatozoa in excess and this contributes to the prevention of polyspermy.

It has been shown that at least one enzyme is activated, within a few seconds, after fertilization. This is a NAD kinase, the enzyme which phosphorylates the respiratory co-enzyme NAD (nicotinamide adenine dinucleotide) into its phosphorylated form, NADP. This synthesis or activation of NAD kinase is the fastest biochemical change that has so far been detected when sea urchin eggs are fertilized.

When sperm is added to sea urchin eggs, there is a big burst in CO_2 production, much larger than could be accounted for by cell oxidation. This burst is due to the production of a "fertilization acid", as Runnström first found in 1928. For many years, numerous unsuccessful attempts have been made to identify its chemical nature. The mystery was solved when Epel discovered, in 1976, that a proton efflux takes place at fertilization; this efflux acidifies the seawater which surrounds the eggs; as a result a burst of CO_2, at the expenses of carbonates and bicarbonates present in sea-water, takes place. Thus, the Na^+/H^+ exchange which, as we have seen, occurs soon after fertilization has two consequences: increase of the internal pH of the egg and production of a CO_2 burst.

A few words about the changes in respiration after fertilization in *frog eggs*. Their oxygen consumption is not affected, but there is, as in the sea urchin eggs, an initial burst in CO_2 production. This excess CO_2 has a different origin than in sea urchin eggs. When the unfertilized eggs are accumulated in the uterus of the female frog, CO_2 cannot escape through the mucus-coated uterine walls. The unfertilized eggs, at the time of laying, are thus intoxicated by a large excess of CO_2. The true CO_2/O_2 ratio (respiratory quotient: R.Q.) drops from the value 1, characteristic of carbohydrate oxidation found in the unfertilized egg, to the very low value of 0.65 after fertilization. This low R.Q. suggests that substrate oxidation is incomplete soon after fertilization and during early cleavage in frog eggs.

3.2.2 Carbohydrate Metabolism

Both sea urchin and frog eggs contain all the enzymes needed for the classical metabolic pathways of carbohydrate metabolism (glycolysis, tricarboxylic acid cycle and hexose-monophosphate direct oxidation). In sea urchin eggs, it has been reported that the glycogen content decreases during the first 10 min after fertilization and then slowly returns to its initial value.

3.2.3 Protein Metabolism

Protein synthesis will be discussed later in a special paragraph. Fertilization in the sea urchin induces a number of changes in the *solubility* and in the *electrical charge* of the proteins; the electrophoretic pattern changes and a new KCl-soluble protein can be isolated. This protein, which is localized in the egg cortex and contains –SH (sulhydryl) groups, probably plays a role in the cortical changes which follow fertilization and lead to the first cleavage. The sulfhydryl–disulfide (–SH/–SS–) ratio undergoes changes in this KCl-soluble protein fraction after fertilization; these changes are probably related to the contractility of the cortex, since this protein presents all the characteristics of the "contractile" proteins. It will be discussed in more detail in the next chapter.

A temporary decrease in total proteins and particularly in the ribosomal proteins has been observed soon after fertilization of the sea urchin egg. Probably correlated with this proteolysis is a transient activation of several distinct proteases, which has been reported by several authors and which, as we shall see, might be physiologically important.

3.2.4 Lipid Metabolism

As for glycogen and proteins, fertilization of the sea urchin egg is followed by the breakdown of some of the lipid constituents. The content in *phosphatides* decreases, while the *free cholesterol* increases at the expense of the cholesterol esters. Of some interest is the fact that the sperm contains a lecithinase, an enzyme which transforms the phosphatides in the surface active lysolecithins.

4 Nucleic Acid and Protein Synthesis

A sudden and important burst of protein and DNA synthesis, in the absence of appreciable RNA synthesis, makes sea urchin egg fertilization one of the most fascinating problems of molecular embryology. We shall first examine DNA synthesis.

4.1 DNA Synthesis

DNA synthesis begins, in the two pronuclei, a few minutes after insemination. It is complete, in sea urchin eggs, in about 20 min, thus before amphimixy. The DNA content of each of the pronuclei exactly doubles, so that the DNA content of the zygote nuclei is four times that of a haploid cell (4 C). In amphibian eggs, the microinjection of labelled *thymidine* (the classical precursor used in the study of DNA synthesis) shows that replication of DNA begins shortly after fertilization and is complete before amphimixy.

We have seen, when we studied maturation, that the whole machinery for DNA synthesis is made ready for use during that period. In particular, DNA polymerase α is synthesized during maturation. This enzyme is present in the cytoplasm, in an active form, since foreign DNA is replicated if it is injected into the egg. However, the DNA present in the egg chromosomes is not replicated until a few minutes after fertilization; DNA synthesis does not begin in the male pronucleus before it swells. In cases of polyspermy, DNA synthesis takes place simultaneously in all the sperm nuclei and in the female pronucleus. Its initiation thus results from *cytoplasmic* changes, perhaps from a migration of cytoplasmic DNA polymerase into the nuclei. As we have seen, initiation of DNA synthesis is not related to the cortical reaction and to Ca^{2+} release since it takes place in ammonia-treated unfertilized eggs, where the Ca^{2+}-dependent step is bypassed.

It is likely that the cytoplasmic changes which induce DNA synthesis are correlated with the rise in internal pH which follows fertilization. There is no doubt that DNA polymerase α is involved in the initiation of DNA synthesis in fertilized or activated sea urchin eggs: a specific inhibitor of this enzyme, *aphidicolin* stops development at amphimixy and completely prevents further DNA replication (Fig. 35).

4.2 RNA Synthesis

There is no sudden burst in RNA synthesis at fertilization. Very little, if any, nuclear RNA synthesis can be found before the two- to four-cell stage; the low level RNA synthesis, which can be detected in recently fertilized sea urchin eggs, is mainly due to mitochondrial RNA synthesis. Suppression of RNA synthesis by actinomycin D treatment has no effect at all on fertilization or even cleavage in both sea urchin and amphibian eggs.

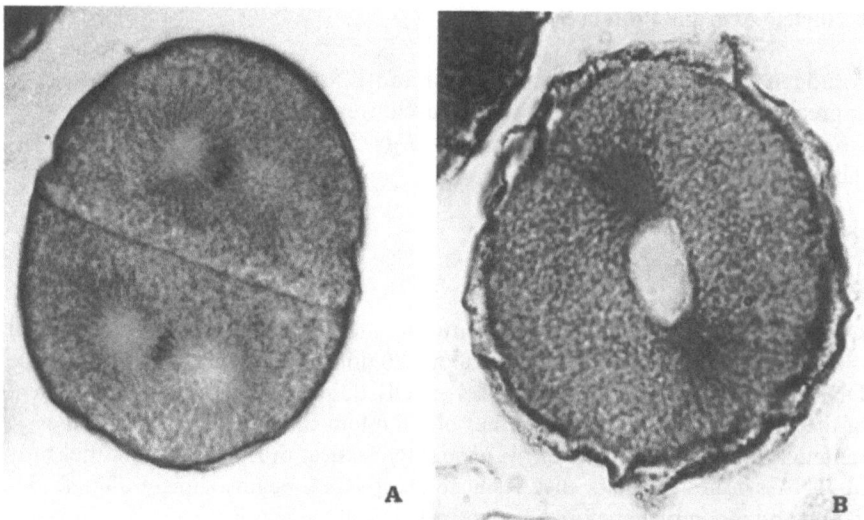

Fig. 35 A, B. Effects of aphidicolin on fertilized sea urchin eggs. **A** Control 2-cell stage embryo in mitosis. **B** Uncleaved aphidicolin-treated egg

This lack of RNA synthesis is probably somehow correlated with the presence of a large store of many kinds of RNAs which, as we have seen, have accumulated during oogenesis. As we know, unfertilized sea urchin eggs possess a large population of mRNAs theoretically capable of coding for as many as 10,000 to 20,000 different proteins. Curiously, maternal mRNAs in sea urchin and *Xenopus* eggs seem to be unfinished molecules: they contain repetitive DNA transcripts and are more similar to nuclear RNAs than to typical mRNAs (E. Davidson). These maternal mRNAs are bound to proteins in the form of *ribonucleoprotein particles* (RNP) smaller than the ribosomes; more will be said about them later.

Although there is almost no net RNA synthesis at fertilization, addition of sperm or activation with ammonia are quickly followed by *polyadenylation* of pre-existing mRNAs: the average length of the (poly A) "tail" increases from 100 to 200 adenylic acid residues. Polyadenylation presumably increases the stability of the maternal mRNAs in the egg cytoplasm. However, the biological significance of polyadenylation at fertilization remains obscure: it can be inhibited by cordycepin, an analogue of adenosine; yet, fertilization and cleavage are normal in eggs treated with this drug.

It is probable that in sea urchin eggs, the excess of ribosomes and mRNAs is stored in *heavy bodies:* these large, RNA-rich granules appear in the cytoplasm after germinal vesicle breakdown and disappear during early cleavage. They increase in size whenever cleavage is inhibited. The identity of the RNAs present in the heavy bodies remains uncertain, but recent work in our laboratory has shown that they contain large amounts of ribosomal RNA.

4.3 Protein Synthesis

Monroy's experiments were the first to demonstrate that fertilization stimulates, in sea urchin eggs, protein synthesis in the same dramatic way as oxygen consumption (Fig. 36). In order to circumvent the impermeability barrier of the unfertilized eggs, he injected a labelled amino acid (methionine) into female sea urchins. The amino acid penetrated very well into the mature ovarian oocytes, but was not incorporated into the proteins of the unfertilized eggs. A few minutes after sperm addition, the free labelled amino acid began to be very actively incorporated into the proteins, demonstrating that intense protein synthesis was now under way.

The first explanation proposed for this sudden burst in protein synthesis was the following: the pronuclei, in the fertilized eggs, would immediately produce new mRNA molecules, which would combine with the pre-existing ribosomes; the newly formed polysomes would be the site of protein synthesis. But we have just seen that there is no measurable burst of RNA synthesis at fertilization and that no nuclear RNA synthesis can be detected until early cleavage. Furthermore, actinomycin D, which should block RNA synthesis, does not prevent the rise of protein synthesis at fertilization.

However, it was still possible that these negative results were simply due to a low permeability of the fertilized eggs for the RNA precursors and actinomycin. This possibility has been ruled out by experiments made on anucleate frag-

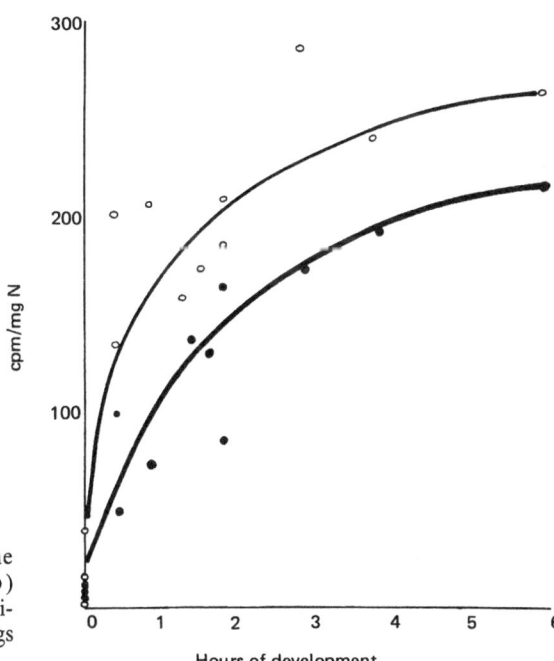

Fig. 36. Fast labelling of the microsomal (○) and soluble (●) protein fractions following fertilization of *Paracentrotus* eggs (Monroy 1960)

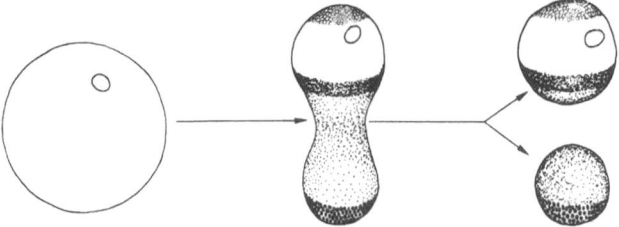

Fig. 37. Centrifugation of unfertilized sea urchin eggs results in the sedimentation of the various inclusions present in the egg. Finally, the eggs break into two parts; only the lighter one contains a nucleus

ments of sea urchin eggs (Fig. 37). The potentialities for development and macromolecular synthesis of such fragments will be examined in another chapter. For our present purpose, it is enough to say that unfertilized eggs of a certain sea urchin species, *Arbacia,* can easily be cut into two halves, nucleate and anucleate, by centrifugation under proper experimental conditions. Neither nucleate nor anucleate fragments of unfertilized eggs are capable of measurable protein synthesis. But, if these fragments, are *activated* by treatment with hypertonic seawater – which induces the cortical reaction, with formation of a fertilization membrane in the egg fragments, but does not allow further development – both nucleate and anucleate halves become capable of active protein synthesis. In fact, incorporation of amino acids into proteins is higher in the *anucleate* than in the nucleate halves. That the anucleate halves, after activation, are the site of true protein synthesis is proven by the fact that the incorporation of amino acid is inhibited all classical inhibitors of protein synthesis. Furthermore, physical methods show that anucleate activated fragments and normally fertilized eggs synthesize the same proteins (in particular, tubulin).Finally, it has been demonstrated that the anucleate fragments contain polysomes, which are actively engaged in protein synthesis. Active polysomes cannot be detected in anucleate fragments which have not been activated by hypertonic seawater treatment.

These experiments definitely rule out the possibility that the stimulation of protein synthesis which follows fertilization is due to the synthesis of mRNAs by the pronuclei. An identical stimulation can be obtained in the complete absence of the nucleus. This situation reminds us of what has been said in the preceding chapter about maturation in amphibian eggs. We saw that progesterone induces the synthesis of the same proteins, whether the germinal vesicle is present or has been surgically removed. In both cases, protein synthesis is necessarily controlled at the level of *translation* and not of transcription.

There are two possible explanations left for the stimulation of protein synthesis in fertilized sea urchin eggs and in activated anucleate fragments. One of them is the existence of *masked mRNA* molecules, which would have been synthesized beforehand by the oocyte and would be stored in the unfertilized egg; these mRNAs would be unable to bind to ribosomes to form polysomes and to support protein synthesis. The second possible explanation is that mRNAs are synthesized on a *cytoplasmic DNA* template after activation or fertilization. Such newly formed cytoplasmic mRNAs would bind to the pre-existing ribosomes and direct protein synthesis. It should be added that these two possibilities are not necessarily mutually exclusive.

Let us begin with the second possibility, that of an intervention of *cytoplasmic DNA* in protein synthesis after fertilization or activation. It is a fact that the nucleate and anucleate halves of the unfertilized sea urchin eggs (which contain about ten times more DNA than the spermatozoon) have about the same DNA content. This proves that the major part of the DNA present in sea urchin eggs is localized in the cytoplasm.

This cytoplasmic DNA is, as expected for mitochondrial DNA, made of circular molecules. It can be transcribed, and there is indeed a low level of RNA synthesis in activated anucleate fragments of sea urchin eggs. But all attempts to demonstrate that mitochondrial RNA can bind to cytoplasmic ribosomes and direct the synthesis of a large number of proteins have failed. The number of proteins synthesized by sea urchin eggs exceeds 400; the information contained in mitochondrial DNA would allow, at best, the synthesis of about a dozen different proteins. It should be added that unfertilized sea urchin eggs synthesize, but at a very low rate, the same set of 400 proteins as the fertilized eggs. It is thus the *rate* of protein synthesis which increases dramatically at fertilization and there are no major qualitative changes in protein synthesis at that time.

We can conclude from this analysis that protein synthesis directed by cytoplasmic, mitochondrial DNA and RNAs probably occurs in the fertilized sea urchin egg, but that it has a minor importance and is probably limited, at best, to the synthesis of a few mitochondrial proteins. The bulk of the proteins synthesized at fertilization must have another origin; the only possibility is the intervention of *masked mRNA molecules of maternal origin.*

What is the evidence in favor of this masked mRNA hypothesis? Monroy has shown and Spiegelman has confirmed that unfertilized sea urchin eggs contain large amounts of "template" RNA. This means that if RNA extracted from unfertilized eggs is added to ribosomes isolated from bacteria or mammalian liver, it supports protein synthesis in this in vitro, acellular system. In order to support protein synthesis, it must bind to the bacterial or liver ribosomes and form polysomes. This is, of course, one of the fundamental properties of the mRNAs. There is no big change in the template activity of the RNAs extracted from sea urchin unfertilized eggs and larvae (*plutei*); the same is true when one compares the template activity of the RNAs isolated from unfertilized frog eggs and tadpoles. This means that the amount of template RNA present in the unfertilized eggs is high enough to support, at least in theory, the intense protein synthesis which exists in plutei and tadpoles. That this template RNA is a mixture of true mRNAs has been demonstrated in Gross' laboratory. It could be shown that unfertilized sea urchin eggs (and even anucleate fragments) contain, in masked form, the specific mRNAs for tubulin, actin and histones.

The masked mRNAs or template RNA present in the unfertilized eggs probably originate in the nucleus of the oocyte during oogenesis. We have seen that about 3% of the RNAs synthesized at the lampbrush stage by the amphibian oocyte is of the mRNA type. It is likely that a large part of the mRNAs synthesized during oogenesis is *stored* in an inactive form and is used at later stages of development only; their utilization begins at fertilization. As already mentioned, the maternal mRNAs are bound to proteins in the form of 60 S ribonucleoprotein (RNP) particles smaller than the ribosomes. The mRNAs associated with these particles cannot be translated in an in vitro system unless the protein moiety of the RNP particles has been removed. Evidently, the maternal mRNAs of the unfertilized sea urchin eggs are in a *masked,* inactive form so long as they are bound to the proteins of the RNP particles.

How are these maternal mRNAs *unmasked* at fertilization? It is only known with certainty that the stimulation of protein synthesis, in sea urchin eggs at least, is closely linked to the *increase in internal pH* which takes place shortly after fertilization. This has been shown by experiments in which the internal pH value has been manipulated by addition of either ammonia or acetate to the eggs. While ammonia, as we have seen, stimulates both DNA and protein synthesis, acetate, which lowers the internal pH of the eggs, inhibits protein synthesis and arrests development at amphimixy (like the classical and potent inhibitors of protein synthesis, puromycin and emetine). There is thus some evidence for the view that, at least in sea urchin eggs, unmasking of maternal mRNAs is a pH-dependent process. But there are so many pH-dependent processes occurring constantly in living cells that such a statement does not lead us very far. More work is obviously needed for a better understanding of the changes that the protein-synthesizing machinery undergoes at fertilization.

One possibility, which was proposed by Monroy in 1971, is that the *ribosomes* of unfertilized sea urchin eggs might be abnormal in the sense that they would be unable to bind maternal mRNAs; mild digestion with trypsin indeed restores the ability of unfertilized egg ribosomes to bind to an artificial messenger, polyuridylic acid (poly U). Since there is some evidence for the release of a protease at fertilization, one can imagine that this enzyme removes some inhibitory proteins from unfertilized egg ribosomes; this would allow these ribosomes to bind more efficiently to the maternal mRNAs. Such a protease could also play a role in the unmasking of the mRNAs stored in the RNP particles. It is only to be expected that the activity of this proteolytic enzyme is pH- dependent. However, one must admit that the data, which have accumulated on the possible differences between the ribosomes of unfertilized and fertilized eggs, are contradictory. A recent publication reported that there is a 15 min lag before ribosomes from unfertilized sea urchin eggs become active in in vitro protein synthesis, while there is no lag for ribosomes from fertilized eggs. It has also been shown that, a few minutes after fertilization, one of the ribosomal proteins (called S6) is phosphorylated. Since phosphorylation of this S6 protein also takes place during *Xenopus* oocyte maturation and when mammalian cells cultured in vitro are induced to multiply at a faster rate, it is likely that this phosphorylation process has some biological importance. But there is no evidence so far that phosphorylation of the S6 ribosomal protein affects the rate of protein synthesis.

It seems therefore unlikely that changes in ribosome conformation or composition, resulting from phosphorylation or mild proteolysis of the ribosomal proteins, are the *only* factors involved in the stimulation of protein synthesis at fertilization.

Evidently many *control mechanisms* operate when the rate of protein synthesis increases sharply at sea urchin egg fertilization. Negative, inhibitory controls must exist in the unfertilized eggs in order to keep protein synthesis at a very low level. However, the suppression of protein synthesis in these eggs is not complete: they already contain a few polysomes and, as we have seen, they synthesize trace amounts of hundreds of proteins. Fertilization, probably by raising the internal pH of the eggs, suppresses the negative controls which were operating in the unfertilized eggs. Thus, its main biochemical effect is to increase the availability of translatable maternal mRNAs and to allow them to bind to the ribosomes. Another factor involved in the increase in protein synthesis, which follows insemination, is a faster elongation rate of the growing polypeptide chains.

Can the findings obtained with sea urchin eggs be generalized to other cells and eggs? Very similar facts have been found in the *germinating seeds* of many plants: the ribosomes are inactive in the dry seeds, which contain, like sea urchin eggs, a store of masked mRNAs. When the seeds are soaked in water in order to induce germination, the preformed mRNAs bind to the ribosomes; polysomes are formed and protein synthesis begins. The only difference is that the stimulation of protein synthesis takes place later in germinating seeds than in sea urchin eggs: it takes 1.5 to 2 h in seeds, but only 4–5 min in sea urchin eggs before a large increase in the rate of protein synthesis can be observed.

When one examines what is known about protein synthesis in eggs other than those of the sea urchin, one comes to the conclusion that the latter represent an extreme case. Unfertilized eggs of the starfish, the mollusk *Mactra* and the worm *Cerebratulus* are active in protein synthesis; the latter is increased after fertilization, but not in the same dramatic manner as in the sea urchin egg. We have already seen that the sea urchin egg is also a special case insofar as oxygen consumption is concerned. One may wonder whether protein synthesis would not *decrease* after the fertilization of eggs which show a diminution of oxygen consumption at fertilization. Experiments on the polychaete worm *Chaetopterus* have shown us that there is a temporary arrest of protein synthesis when the eggs are activated by the addition of an excess of KCl to the seawater; such a treatment, like fertilization, decreases the oxygen consumption of the eggs by almost 50%. Therefore, it is possible that *energy production*, resulting from the oxidations and phosphorylations in the mitochondria, might be another control mechanism of protein synthesis at the time of fertilization.

Frog eggs also deserve mention. Their protein-synthetic activity decreases as they move through the genital tract (oviduct, uterus) of the female. At fertilization, protein synthesis reverts to a level equivalent to that of the ovarian egg after maturation. But it is not at all certain that this effect on protein synthesis is due to fertilization itself. It could be due to the removal of the excess CO_2, which has accumulated when the eggs were packed in the uterus of their mother, as already mentioned. Since an excess of CO_2 in the eggs should lower their internal pH, it is possible that this factor, as in sea urchin eggs, controls the rate of protein synthesis in frog eggs.

We think that additional work, on eggs from a variety of animal species, is required before one can generalize the hypothesis that internal pH controls the rate of protein synthesis. For instance, one may wonder whether, in activated *Chaetopterus* eggs, in which there is a transient decrease in protein synthesis after activation, there is a concomitant decrease in the internal pH; only new experiments, which are now technically feasible, will provide an answer to this question.

In conclusion, it seems that the magnitude of the increase in both protein synthesis and respiration, which occurs at fertilization, largely depends on the stage of repression which prevails in the unfertilized egg. The main effect of fertilization is to remove the inhibitions which make the unfertilized egg a "sleepy," almost "dormant" cell. In other words, the main effect of fertilization would be a derepression rather than an activation; but this *derepression does not operate, as usual, at the genetic, but at the post-transcriptional level.*

5 Parthenogenetic Activation

The experimental embryologists of the beginning of this century were not as sophisticated as the molecular embryologists of today. They made no distinction between derepression and activation, although, as we shall see, one of them at least was very well aware of the fact that the unfertilized egg is inactive because it is "intoxicated" (we would say, in our modern terminology, repressed). Parthenogenetic stimulation of development after treatment of unfertilized eggs by chemical or physical agents was therefore considered as an activation of the "sleeping" egg.

We shall briefly present here the two main methods of parthenogenesis (which were discovered, early in this century, by Loeb and Bataillon) because they still retain a good deal of interest for molecular embryologists. Loeb was working on sea urchin eggs and discovered that they undergo *activation* when they are treated with *hypertonic seawater*. The fertilization membrane is elevated in the normal fashion, but only a single aster, a "monaster" (of course, of maternal origin) can form. At best, the chromosomes can undergo a few cycles of replication around this monaster; since there is only one aster, no typical mitotic apparatus (see the next chapter for details), with two asters and a spindle, can form. The consequence is that the egg cannot divide and will not develop further. But Loeb found that if the activated eggs are submitted, after a few minutes, to a second treatment, with *butyric acid* this time, a second aster will appear and the egg will develop *parthenogenetically* (that is, without the intervention of the spermatozoon) until the larval pluteus stage is reached. As we have seen, the two asters form around typical centrioles in such eggs.

Bataillon worked on frog eggs and found that they can be activated by pricking them with a clean needle. They form a fertilization membrane and undergo a series of monasterial, abortive division cycles. If, however, the needle is not clean and is contaminated with blood cells, so that a nucleate cell is introduced into the egg at the time of pricking, a second aster will form around the injected cell. Mitosis will then be normal and development will proceed until the tadpole stage. The cell which has been introduced in the unfertilized egg thus contains a "second factor" necessary for the formation of a second aster and further development. Since the introduction of an anucleate mammalian red blood cell had no effect, Bataillon concluded that the second factor must somehow be linked to the cell nucleus and that a "karyocatalysis" is required for parthenogenetic development.

This "karyocatalysis" theory is no longer held, since there is now good evidence that Bataillon's second factor is a *centriole,* not a nucleus. The lack of activity, as a second factor, of mammalian red blood cells is due to the fact that these cells do not possess centrioles. It has been shown that injection of cell extracts enriched in centrioles has the same effects as injection of a nucleate cell, i.e. induction of complete parthenogenetic development. Treatment with colchicine, which inhibits the formation of astral microtubules, suppresses the activity of these cell extracts. However, this activity is enhanced by agents, which, like heavy water (D_2O), stabilize the microtubules. The second factor of Bataillon is thus a *"microtubule organizing center"* (MTOC); it is necessary for the building up of a bipolar mitotic apparatus in eggs which have been activated by pricking with a needle.

It is important to note that parthenogenetic activation, even if it is not followed by further development, induces the same metabolic changes in the egg as fertilization. Treatments with hypertonic seawater, even in the absence of a second treatment with butyric acid, is sufficient to increase both respiration and protein synthesis in sea urchin eggs. The part played by the spermatozoon in these early metabolic events thus appears as negligible.

This statement is no longer true for later development. Parthogenetic embryos are, as a rule, *haploid* and die during development. The *haploid syndrome* can very easily be demonstrated in frog eggs which have been fertilized with sperm previously irradiated for a short time (1 to 2 min) with an UV lamp. The spermatozoa remain perfectly motile and the same percentage of fertilization can be obtained as with normal, nonirradiated sperm. But the chromatin of the spermatozoon has been irreversibly damaged; the paternal chromosomes are eliminated into the cytoplasm, where their DNA undergoes complete degradation, during the first cleavage of the egg. Development is at first normal; but a few eggs already fail to gastrulate properly. The others may reach, at best, a very abnormal tadpole stage. They display strong oedema, have a reduced head and very small gills, and are unable to digest their yolk completely. This haploid syndrome is always *lethal.* A few eggs, however, may escape death and become adults. Cytological observation shows that they are always diploid or polyploid. They contain $2n$, $4n$, $6n$ or $8n$ chromosomes. This regulation of the number of chromosomes, which was first observed by Bataillon with his pricking method, is due to the fact that, in a few eggs, the chromosomes undergo one or more monasterial cycles of replication before a second aster forms around the injected nucleate cell. The biochemical reasons for the haploid syndrome remain totally unknown.

6 Molecular Embryology and Classical Theories of Fertilization

Both Loeb and Bataillon formed theories of fertilization based on their own work on sea urchin and frog eggs, respectively.

Loeb proposed that the two factors needed for inducing parthenogenetic development in sea urchin eggs (hypertonic seawater and butyric acid, or vice versa) could have two opposite effects. Incipient cytolysis would first be induced and would lead to the death of the egg unless restoration to normal conditions was engendered by the second parthenogenetic agent. By analogy, the spermatozoon would, during fertilization, first induce lytic changes and then restore the normal situation. There is some biochemical evidence supporting Loeb's theory, at least in the case of the sea urchin egg. We have mentioned the initial breakdown and later resynthesis of macromolecules such as glycogen, certain proteins, phosphatides and cholesterol esters. The presence of a lecithinase in spermatozoa, as well as the synthesis or activation of proteases very soon after fertilization, can be taken as arguments in favor of the incipient cytolysis hypothesis. However, the breakdown of the cortical granules is a normal response to a local increase in calcium ions; the exocytosis of its contents, can hardly be taken as a lytic process.

For Bataillon, the egg is an intoxicated cell and the elimination of the perivitelline fluid would lead to a "purification" of the cell (*réaction d'épuration*). Although this part of the theory is probably no longer true, since certain eggs show very little, if any, cortical changes at fertilization, molecular embryology provides ample demonstration for the view that the unfertilized egg is intoxicated (or repressed) by the products of its own metabolism. This intoxication is probably related to the very low permeability of the unfertilized egg. As we have seen, the studies made on the respiration of invertebrate eggs have shown that oxygen consumption is abnormal in unfertilized eggs and is brought back to normality by fertilization.

The fact that the cytoplasm of frog oocytes, which have finished their maturation, blocks cell division when it is injected into a cleaving egg is further evidence in favor of Bataillon's theory.

Bataillon has placed emphasis on the role of CO_2 in the intoxication of the unfertilized egg. In fact, he could show that if frog eggs are treated with an excess of CO_2, they undergo activation (monasterial cycles of chromosomes) instead of cleavage. If CO_2 is removed, normal cleavage becomes possible. Some of the experiments described in this chapter have shown that the CO_2 content of the eggs might indeed play a role by controlling the intracellular pH in the regulation of protein synthesis in both sea urchin and amphibian eggs.

When we look today at the theories proposed long ago by Loeb and by Bataillon, we come to the conclusion that these two great embryologists showed remarkable insight. At present, their theories no longer contradict each other, but simply complement each other in a harmonious way.

CHAPTER VI

Egg Cleavage: A Story of Cell Division

1 General Outlook

In Chap. III, we already mentioned that various kinds of eggs undergo different types of cleavage. Depending on the amount and localization of yolk, cleavage will be equal and total, unequal and total, partial or superficial. These differences are, in general, due to the fact that the yolk constitutes a barrier to the progress of the cleavage furrows which divide the egg into its blastomeres. This explanation is not, however, always satisfactory. It does not explain why cleavage is unequal in the eggs of mollusks (see Fig. 12) and worms, since the distribution of the yolk does not show any peculiarity in these eggs. The reasons for this unequal cleavage are not really known; it must depend on some hidden feature in the architecture of the egg, since one can rule out particular properties of the spindle and asters. Since, as we shall see, the egg cortex plays a most important role in the formation of the cleavage furrow, one could reasonably suppose that the molecular architecture of the cortex is asymmetric in eggs which undergo unequal cleavage; but this hypothesis has still to be proven. Genetic factors are also involved: for instance, in the mollusk *Lymnaea,* dextrality or sinistrality of the shell is due to a single gene, with dextrality dominant. The product of this gene during oogenesis affects the cleavage pattern: injection of dextral cytoplasm into uncleaved sinistral eggs changes their cleavage pattern into dextral cleavage. Therefore, although the dextral gene product is synthesized during oogenesis, it does not function before cleavage begins. Unfortunately, we know nothing about its chemical nature.

Whatever the type of cleavage might be, the period of development which immediately follows fertilization is characterized by the speed and frequency of the *cell divisions* (mitoses). Mitosis is not fundamentally different in eggs and in other cells. As shown in Fig. 38, all the classical phases of mitosis can be easily recognized during egg cleavage. The chromosomes become clearly visible at *prophase;* they form an equatorial plate, in the middle of the spindle at *metaphase;* they move toward the two poles of the spindle (where very large asters are present around the centrioles) at *anaphase.* The nuclei reform at *telophase,* when the cleavage furrow separates the cleaving blastomeres into two daughter cells. However, in mouse eggs, there are no asters during the first cleavages.

Mitoses follow each other very quickly in the cleaving egg except in those of the mammals: this high mitotic activity leads to a fragmentation of the egg into

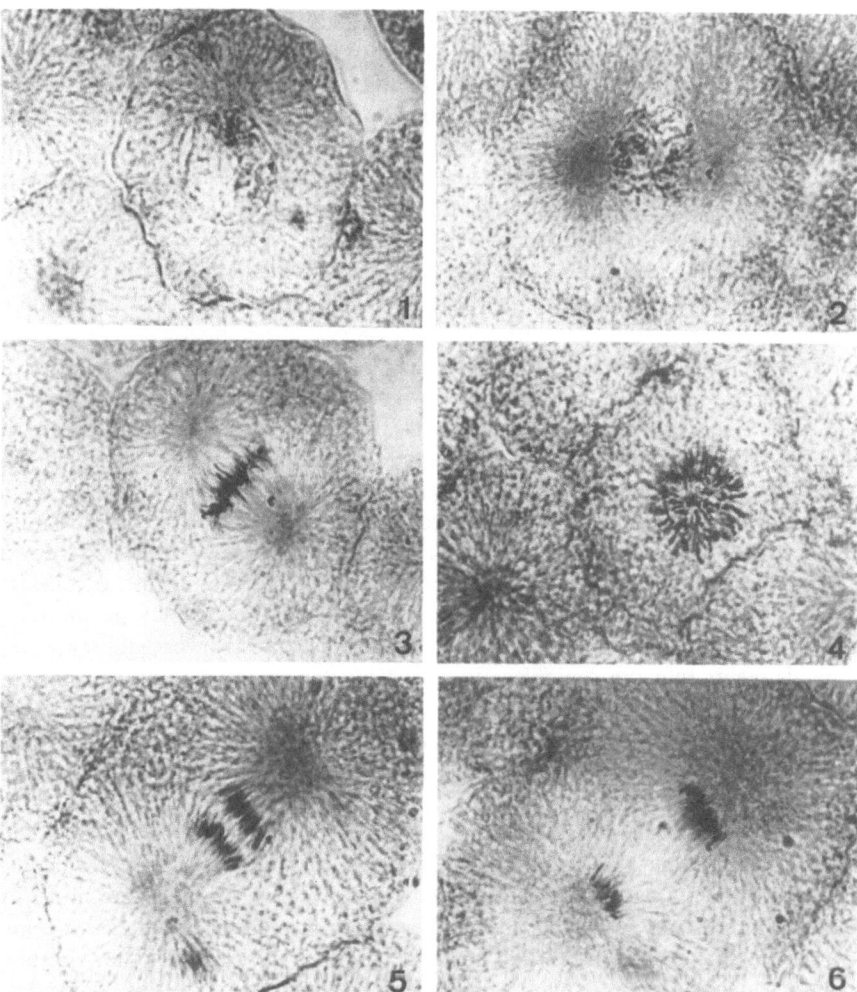

Fig. 38. Mitosis in whitefish blastulae. *1* Interphase nucleus; *2* prophase; *3* metaphase lateral view; *4* metaphase showing the equatorial plate; *5* beginning of anaphase; *6* end of anaphase

an evergrowing number of *blastomeres.* The cleaving egg first reaches the *morula,* then the *blastula* stage, which marks the end of the cleavage period. In many eggs, for instance, those of the sea urchin and the amphibians (Fig. 13), the blastomeres exude a fluid toward the center of the egg. Thus, the blastomeres come to surround a cavity filled with this fluid; it is the *blastocoele,* which contains, besides water and salts, proteins and glycogen secreted by the neighboring cells.

At the beginning of cleavage, the mitoses are perfectly *synchronous;* later on, synchrony is lost. The cells divide faster near the animal pole (where they form the *micromeres*) than close to the yolk-laden vegetal pole (where the *macromeres* are formed; see Fig. 13). The reasons for this loss of synchrony are not well known. Asynchrony is probably not due to differences in the speed of DNA replication in the individual nuclei and is almost certainly under the control of cytoplasmic factors. The amount of yolk in the various blastomeres is certainly one of these cytoplasmic factors. The fact that synchrony is lost faster than usual after slight centrifugation, which modifies the yolk gradient by sedimenting the yolk platelets at the vegetal pole, supports this suggestion.

When cleavage becomes asynchronous, the percentage of dividing cells (mitotic index) drops: for instance, in sea urchins 60% of the cells are in mitosis in a young (6 h after fertilization) blastula, but only 10% are dividing a few hours later, at the time of hatching. Curiously, in sea urchin eggs, there is a *gradient of mitotic activity* from the 16-cell stage onwards. It decreases from the vegetal to the animal pole and it is suppressed by treatment with actinomycin D. Thus, it looks as if transcription of DNA sequences in the micromeres controlled the synchronization of the mitoses during cleavage in sea urchins.

In *Xenopus* (see Fig. 13), 12 synchronous and fast (they take only 30 min) cleavages follow each other after fertilization; mitotic activity then slows down and synthesis of several kinds of RNAs (see later, for details) begins. If cells are isolated from dissociated morulae or early blastulae, they display no motility; dissociated cells become motile when cleavage becomes slower and asynchronous. It seems that this *"midblastula transition"* results from a critical ratio between the volumes of the nucleus and the cytoplasm (nucleocytoplasmic ratio) and it does not depend on the number of cleavages, the number of DNA replication cycles or the time elapsed since fertilization as shown by Newport and Kirschner's (1982) elegant experiments.

A few peculiarities of egg cleavage, as compared to classical cell division, deserve mention. A striking fact should be referred to at once. As a general rule, the nuclei of cleaving eggs have *no nucleoli*. Instead of true nucleoli, one finds, by electron microscopy, nucleolar-like bodies which have been called *cleavage nucleoli*. They have a purely fibrillar structure and lack the classical outer granular region, which is normally rich in ribonucleoprotein particles (see Fig. 20). In fact, cytochemistry does not show the presence of RNA in the cleavage nucleoli; it is also impossible to detect any incorporation of labelled uridine, the classical precursor of RNA synthesis. These phenomena are generally attributed to the fact that these fast dividing nuclei do not have enough time to elaborate true nucleoli. But such an explanation does not seem very plausible. If one slows down or even stops mitosis in cleaving eggs by the addition of inhibitors of DNA synthesis (e.g. hydroxyurea, fluorodeoxyuridine, deoxyadenosine), the number and size of the cleavage nucleoli increase. But they remain typical cleavage nucleoli and do not change into true nucleoli. They contain no RNA and they do not incorporate labelled uridine; they are still fibrillar and devoid of an outer granular zone. Since cleavage nucleoli are made of proteins, they probably represent an accumulation in the nucleus of proteins which have been synthesized in the cytoplasm. What is certain is that cleavage nucleoli can form even under conditions in which RNA synthesis and energy production are almost completely suppressed.

One has to wait until the late blastula stage, just before the gastrulation movement begins, to find true RNA-rich nucleoli in the cells. They make their appearance in all cells, at the vegetal as well as at the animal pole, at about the same time.

There is one noteworthy exception to the rule that cleaving eggs do not have true nucleoli. This is the case of the mammalian eggs, in which large, RNA-rich nucleoli make their appearance; cleavage in these eggs is unusually slow (see Chap. IX).

There is nothing particular about the morphology of the chromosomes in cleaving eggs as one can see in Fig. 38. However, we should mention mitotic abnormalities which are very frequent in fast cleaving eggs as soon as they are placed under slightly adverse conditions.

The frequent occurrence of *cytasters* (Fig. 34) has already been mentioned. Their formation is not linked to the presence of the nucleus, since they appear when anucleate fragments of sea urchin eggs are treated with heavy water (D_2O). We have seen that their production requires the previous mixing of the nuclear sap of the germinal vesicle with the cytoplasm, at the time of maturation. However, the proteins which build up the astral rays (tubulins) are not accumulated in the germinal vesicle, but are dispersed in the cytoplasm. As mentioned previously, cytaster formation does not require DNA, RNA or protein synthesis.

This tendency in the fertilized egg of forming multiple asters is confirmed by the frequent appearance of *pluricentric mitoses,* which lead to unequal distribution of the chromosomes in the daughter cells; the results is *aneuploidy* (the various blastomeres contain different numbers of chromosomes), a condition which is lethal when it is widespread. Development stops at the blastula stage and the embryos soon die. The frequent formation of cytasters and pluricentric mitoses in cleaving eggs shows that their cytoplasm is a particularly favorable environment for the replication of the centrioles. Since this replication is possible in anucleate fragments and in eggs in which the chromosomes have been destroyed by strong X-irradiation, it follows that *achromosomal mitoses* are possible. The egg will divide even when chromosomes are completely lacking. In such cases, the main purpose of mitosis, the equal distribution of the chromosomes in the daughter cells, is obviously defeated.

Less frequent are the *anastral* mitoses, in which a spindle is present, but asters are missing. This is the rule for plant cells, oocyte maturation and early cleaving mammalian eggs, but usually, blastomeres, which contain an anastral mitosis, fail to divide. Another type of mitotic abnormality is sometimes encountered, in which the spindle does not form. Since such abnormal mitoses were first observed after treatment with colchicine, they were given the abbreviated name *c-mitoses.* The chromosomes divide, but remain in the center of the cell. If the latter recovers and if normal spindle and asters are formed, the resulting daughter cells will be *polyploid* (usually tetraploid, with $4n$ chromosomes). C-mitoses are easily obtained, in cleaving eggs, by such simple treatments as heat or cold.

Chromosomal abnormalities are also frequent and result from alterations of the DNA molecules. Chromosomes can break into fragments which do not contain the specialized part needed for the attachment of the chromosomes to the spindle fibers (the *centromere* or *kinetochore*); such fragments are lost in the cytoplasm, where they degenerate. In many instances, and perhaps in all, the kinetochore is made of DNA molecules which are different from the bulk of chromosomal DNA. In the mouse, the highly repetitive AT-rich satellite DNA (see Chap. II) is accumulated in the centromere region.

When chromatin is irreversibly injured and DNA undergoes degradation, it changes into a dense, spherical mass. This is *pycnosis,* a clear sign of cell death. It is noteworthy that pycnosis (and thus cell death) becomes a physiological phenomenon during embryonic development, at later stages than cleavage. Pycnoses are frequent in the eye and brain of perfectly normal young tadpoles, for instance. It seems that, in these organs, a certain number of cells are "sacrificed" for the others, which are able to reutilize the breakdown products of the macromolecules which were present in the dying cells. Extensive cell death, at very precise stages of development and in well-defined regions of the embryo, is a normal process which has been called "programmed cell death". It occurs on a very large scale during metamorphosis in insects and, during tail regression in frog tadpoles. Programmed cell death is also required for the separation of our digits from one another; if it did not take place in our "limb buds", when we were young embryos, we would have webs, like ducks, instead of hands. In many instances, programmed cell death is a response to a hormonal stimulation; in others, it results from a gene mutation. Its biochemical mechanism is not well understood: it has been suggested that cell death results from the release in the cytoplasm of the hydrolytic enzymes present in the lysosomes, but there is no doubt that more complex mechanisms are involved: for instance, regression of a tadpole's tail absolutely requires RNA and protein synthesis.

Finally, it is possible, in cleaving eggs, to stop the formation of the *furrows* without interfering with nuclear division. The chromosomes and asters divide repeatedly, so that several daughter nuclei are formed in an undivided egg or a large blastomere. Narcotics, like urethane or ether, are particularly active in stopping furrow formation without affecting chromosome replication, but the most powerful inhibitor of furrowing, in eggs as in other cells, is *cytochalasin,* a drug which prevents actin polymerization.

2 The Biochemistry of Cleavage

2.1 Energy Production

Oxygen consumption increases progressively, in both sea urchin and amphibian eggs, during cleavage. This increase seems to be controlled in the usual way by the concentration of the main phosphate acceptor, ADP (adenosine diphosphate) in the egg.

On the other hand, the effects of *anaerobic conditions* (absence of oxygen) and of *cyanide* (which inhibits respiration by combining with cytochrome oxidase) are different in sea urchin and amphibian eggs. While cleavage is almost immediately arrested in the former, frog eggs slowly reach an early blastula stage in the total absence of oxygen or in the presence of cyanide, at concentrations which almost completely inhibit oxygen consumption.

But this difference between amphibian and sea urchin eggs is more apparent than real. We know that the true source of energy in living organisms is ATP (adenosine triphosphate) and that the synthesis of ATP from ADP and inorganic phosphate (phosphorylation) is coupled with the oxidations catalyzed by the respiratory chain. This coupling can be interrupted by a poison, *dinitrophenol;* in its presence, respiration increases and the ATP concentration decreases. Despite the rise in respiration, the dinitrophenol-treated cells have less energy available for the synthetic process than normal ones, because they contain less ATP. Now, dinitrophenol blocks cleavage almost immediately in amphibian as well as in sea urchin eggs. Further proof that energy must be supplied in the form of ATP is given by experiments in which the ATP content of eggs placed in the absence of oxygen has been measured. The ATP content drops in both cases, but much faster in the sea urchin than in the frog. Cleavage stops, in both kinds of eggs, when the ATP content has fallen to about 50% of its initial value.

Are there *cyclic variations in oxygen consumption* during cell division? Very careful studies have been made by Zeuthen, who succeeded in measuring, with exceedingly sensitive methods, the respiration of a single sea urchin or amphibian egg during its first cleavages. As shown in Fig. 39, small cyclic variations in oxygen uptake are observed; since such variations are not found when the respiration of unfertilized eggs is measured, it can safely be concluded that the cyclic changes detected during cleavage are real, and not due to experimental errors. In fact, similar changes have been found when the oxidation of radioactive glucose, added to cleaving sea urchin eggs, has been followed. It seems that the peaks of respiratory activity correspond to those stages of mitosis during which the nucleus is surrounded by a membrane (end of telophase, interphase, beginning of prophase). Respiration would be low, on the contrary, during metaphase and anaphase.

2.2 Chemical Nature of the Mitotic Apparatus

The mitotic apparatus comprises the spindle, the chromosomes, the centrioles and the two asters. Cytochemical studies show that it is made of proteins, RNA and glycogen. Electron microscopy demonstrates that the spindle and asters are made of *microtubules* (Fig. 40). The so-called astral rays are, in fact, hollow cylinders. The RNA present in the spindle is simply an accumulation of ribosomes trapped between the microtubules; the same situation prevails for glycogen particles.

Mazia and Dan succeeded in *isolating* the whole, intact mitotic apparatus from cleaving sea urchin eggs (Fig. 41). The mitotic apparatus, except for the

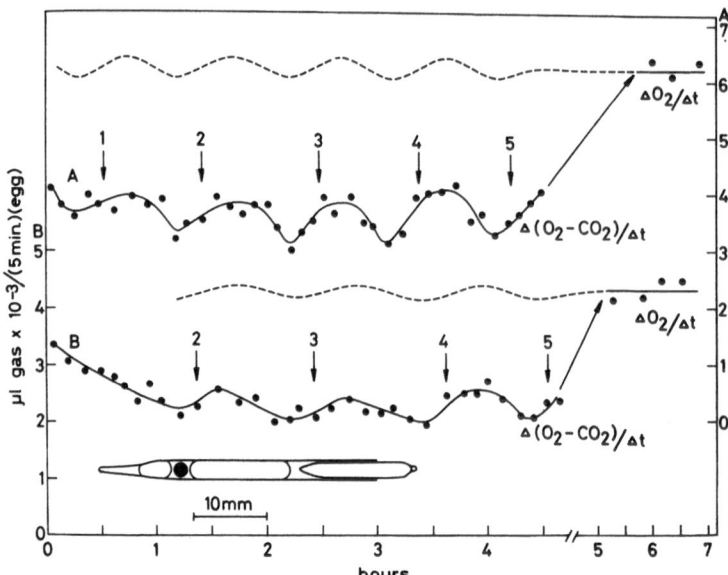

Fig. 39. Rates of gaseous exchanges in two single frog eggs (*A, B*). *Fully drawn curves* show $\Delta\,(O_2 - CO_2)/\Delta t$ and $\Delta O_2/\Delta t$ measured, respectively in the intervals 0–5 and 5–7 h. The *dashed lines* show extrapolated values for $\Delta O_2/\Delta t$. *Arrows* indicate when cleavages *1* to *5* begin. Below is the diver used for the measurements (Zeuthen and Hamburger 1972)

chromosomes and contaminants, such as ribosomes, is made of a variety of proteins; the major one is *tubulin,* the main constituent of microtubules in both interphase and mitotic cells (where they form part of the cytoskeleton and the mitotic apparatus, respectively). Tubulin is a globular molecule made of two nonidentical subunits, and it can be obtained in pure form from pig brain. Under proper in vitro conditions, pig brain tubulin undergoes linear polymerization and forms typical microtubules. In vitro tubulin polymerization is enhanced by addition of other proteins, the *microtubule-associated proteins* (MAPs); in vivo MAPs are associated with tubulin in the mitotic spindle and asters. In the living, dividing cell, microtubule polymerization starts from organizing centers (MTOCs), which act like seeds in the growth of crystals. The main MTOCs are the centrioles (or rather the *pericentriolar clouds* which diffusely surround the centrioles) and the *kinetochores* (centromeres) of the chromosomes. The microtubules present in the mitotic apparatus have thus a double origin: the poles, where the centrioles are located, and the chromosomes. It is believed that polymerization of tubulin starts at these two MTOCs and that it progresses by "treadmilling". New tubulin molecules are added at the end of the growing microtubules; others are lost at the opposite end. At anaphase, depolymerization of the microtubules takes place in the central region of the spindle. It is believed that

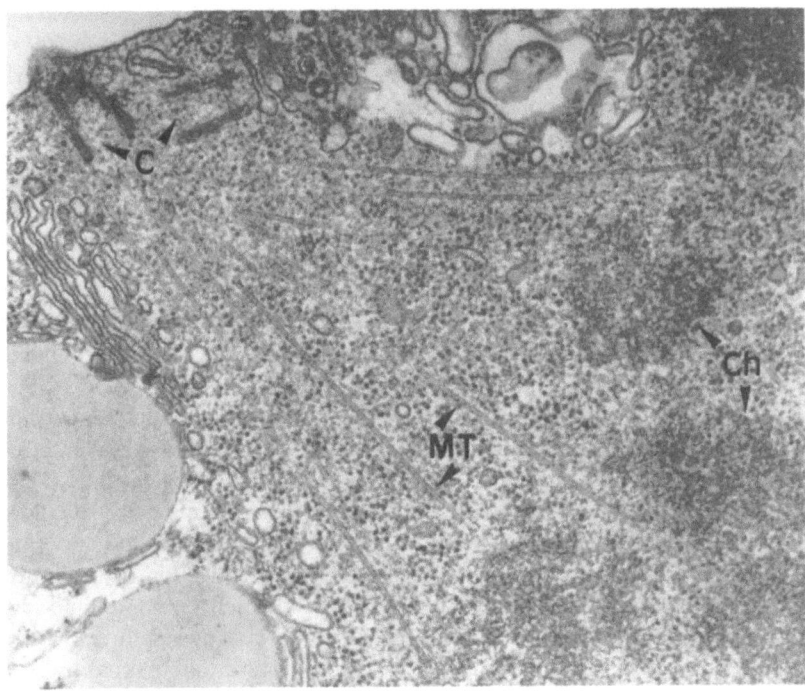

Fig. 40. Mitosis in embryonic sponge cells. *C* centriole: *Ch* chromosome; *MT* microtubule (electron microscopy). (Courtesy of Dr. L. DeVos)

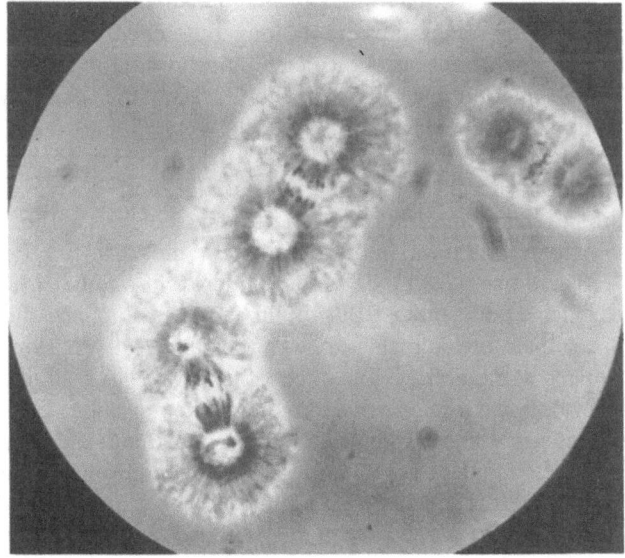

Fig. 41. Mitotic apparatus isolated from cleaving sea urchin eggs (phase contrast microscopy). (Courtesy of Dr. D. Mazia)

this local depolymerization of the microtubules at the equator of the dividing cell results from a local increase in the free Ca^{2+} concentration, but it must be admitted that, despite many years of efforts, the molecular mechanisms underlying the dynamics of cell division remain poorly understood. For instance, some believe that anaphase movement of the chromosomes results from an ATP-driven sliding of the microtubules (as for ciliary beating) which would shorten the half-spindles; others think that the spindle contains contractile proteins similar to actomyosin in muscle. Their ATP-dependent contraction would draw the kinetochores and then the whole chromosomes toward the poles; still others, as just mentioned, propose that anaphase chromosome movement is due to a Ca^{2+} induced depolymerization of the microtubules, starting at the spindle's equator and progressing toward the poles. Several molecular mechanisms might of course co-operate in chromosome separation and movement.

A number of drugs interfere with the building up and the maintenance of the mitotic apparatus. Elegant experiments by Mazia on cleaving sea urchin eggs have shown that reducing agents like *mercaptoethanol* ($HSCH_2-CH_2OH$) dissolve the mitotic apparatus both in the living egg and after in vitro isolation. Interestingly, mercaptoethanol inhibits the formation of the mitotic apparatus in living sea urchin eggs, without interfering with the replication of the centrioles and chromosomes. As a result, eggs which had been left for some time in mercaptoethanol and were then returned to normal seawater cleave directly into four blastomeres, bypassing division into two cells. Concentrated urea and formamide, which break hydrogen bonds, also destroy the mitotic apparatus and stop mitosis in both sea urchin and amphibian eggs.

The classical agent for preventing spindle formation and for stopping mitosis is *colchicine*. We have already mentioned the existence of so-called c-mitoses where the chromosomes are no longer attached to the spindle and become scattered in the cytoplasm. Colchicine binds to tubulin and this prevents in vitro and in vivo microtubule polymerization. Many other drugs (vinblastin, nocodazole, podophyllin, etc.) have similar effects; some of them are extensively used for cancer chemotherapy. In eggs, all these drugs quickly arrest cleavage (Fig. 42) without decreasing the total content in tubulin: both in vivo and in vitro they impede the linear polymerization of the tubulin molecules. The latter accumulate in an unpolymerized form, in the cytoplasm of the colchicine-treated eggs.

There has recently been some interest in another drug, *taxol*. In contrast to colchicine and similar anti-tubulin agents, it stabilizes the existing microtubules. Since their breakdown is now impossible, cell division and egg cleavage are stopped at metaphase. The effect of taxol on eggs is similar to that of an injection of cytoplasm from an unfertilized egg into a cleaving egg: as we have seen when we discussed maturation, this cytoplasm contains a *cytostatic factor* (CSF) which prevents the depolymerization of the meiotic and mitotic spindles. Similarly, taxol "freezes" already existing mitotic apparatuses; in addition, like heavy water, which also stabilizes the microtubules, it induces the appearance of numerous cytasters around amorphous cytoplasmic areas, devoid of well-organized centrioles, in both maturing and fertilized amphibian eggs (Fig. 34).

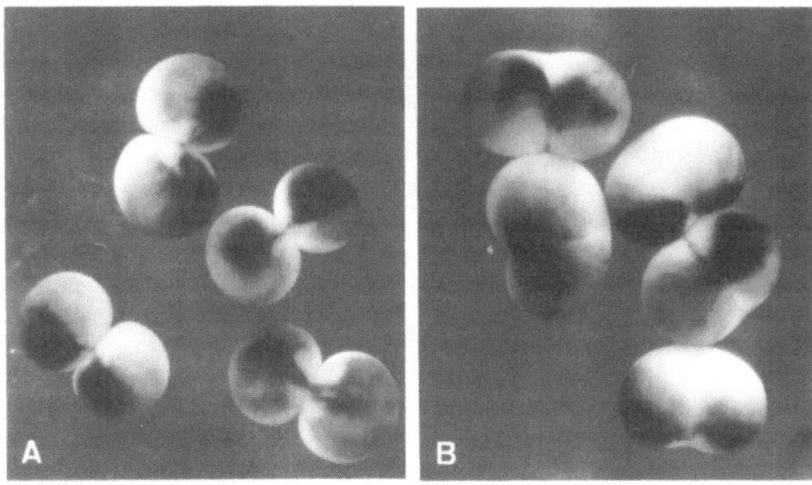

Fig. 42 A, B. Effect of vinblastin on the first cleavage of *Xenopus* eggs. Control (**A**) and vinblastin-treated (**B**) eggs. (Courtesy of Dr. R. Tencer)

2.3 DNA Replication During Cleavage

The first demonstration that extensive DNA synthesis takes place during sea urchin egg cleavage was given by the senior author about 50 years ago. As shown in Fig. 43, there is a close correlation between DNA synthesis and the mitotic index; the latter drops progressively during cleavage, as already mentioned.

There are a number of peculiarities about DNA synthesis in cleaving eggs as compared to somatic dividing cells: in all cells, DNA synthesis occurs during a particular period of the interphase, the so-called S-period. Just after cell division, the nucleus contains the diploid (2 C) content of DNA. No DNA synthesis occurs during the following few hours, called the G_1 period. A schema of the *cell cycle* is given in Fig. 44. The S-period then follows, during which the DNA content exactly doubles, reaching 4 C values (that is, four times the value one finds in the nucleus of a spermatozoon). The S-phase lasts a few hours and is followed by a second gap, the G_2 period. The cell does not immediately divide after DNA replication; other events which are required for entry in mitosis thus occur in G_2, a period which also lasts for a few hours. Then the cell divides and the cycle starts again. Certain cells, of course, never divide anymore; for instance, the nerve cells (neurones) of our brain; they are said to remain in G_0 since they are no longer cycling.

There is one major difference between cleaving eggs and other cells. Interphase is shorter, so that the G_1 and G_2 *periods are practically absent.* In other words, the cleaving egg replicates its DNA "at full speed," just as bacteria do. Only when the egg is actually in mitosis (metaphase, anaphase) does DNA syn-

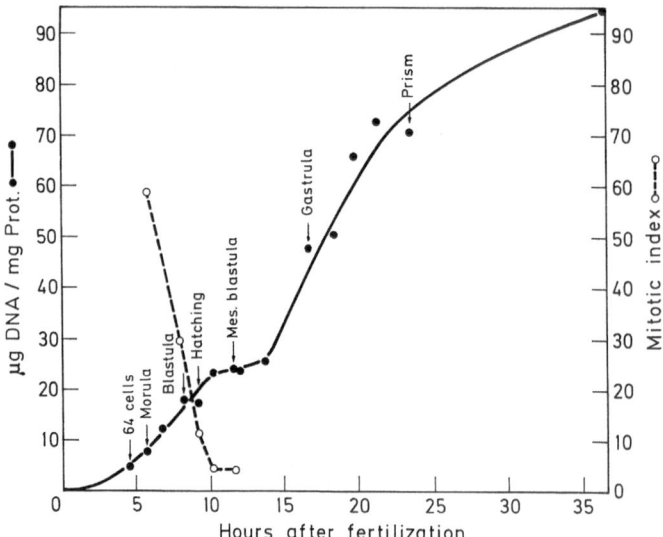

Fig. 43. Increase in DNA content (*solid line*) and drop in mitotic index between morula and mesenchyme blastula (*dotted line*) during sea urchin egg development. (From Parisi et al. 1978, according to Brachet 1933)

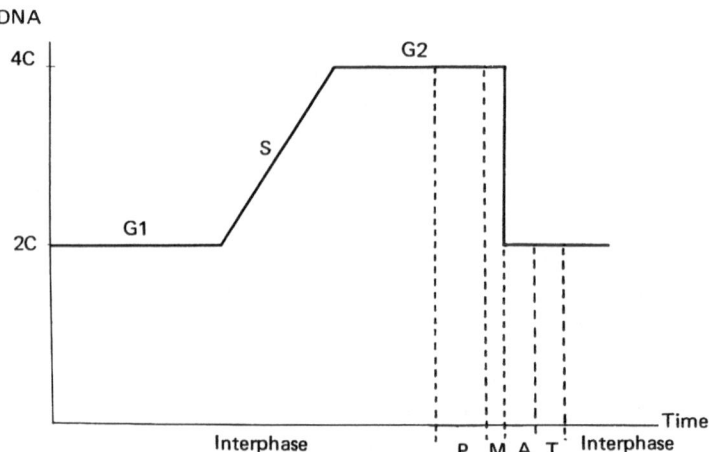

Fig. 44. DNA replication during the S-phase of the cell cycle. *P* prophase; *M* metaphase; *A* anaphase; *T* telophase. (Courtesy of Prof. P. Van Gansen)

thesis stop. Later on, at the end of cleavage, mitotic activity decreases and a typical cell cycle becomes detectable.

In cleaving eggs (except those of mammals), the S-phase is exceptionally short: in large amphibian eggs, the whole genome (total nuclear DNA) is replicated in less than 10 min; the fastest rate of DNA replication known today is that of the *Drosophila*, in which the S-phase lasts only 3–4 min during early cleavages. In contrast, the S-phase lasts several hours in somatic cells and in mouse eggs. Some experiments suggest that the speed of the first cleavages in amphibian eggs is determined by *cytoplasmic* factors and is not an autonomous property of the nuclei: injection of cytoplasm from a fast-cleaving egg into an egg which has a slow cleavage rate speeds up its cleavage (and vice versa).

The reason why the cleaving egg is capable of synthesizing its DNA so fast has already been mentioned. As we have seen, the whole machinery for DNA synthesis is ready in the cytoplasm since maturation took place. The precursors for DNA synthesis, as well as the enzymes needed for its replication, are already present. DNA polymerase α, the enzyme which builds up a new polynucleotide chain on the DNA template, is already present in large amounts in the cytoplasm; at each cell division, it moves from the cytoplasm into the nucleus. Therefore, the total amount of DNA polymerase α is about the same in an unfertilized egg as in a blastula, which contains several hundreds of cells; but the amount of this enzyme is the same, in a *nucleus,* at all stages. The zygote nucleus and one single blastula nucleus have the same content in DNA polymerase. Other cells are not so fortunate as the cleaving eggs. They must synthesize the polymerase which is needed for DNA replication; this takes time and explains the necessity for the G_1 period. But, even in sea urchins, the whole machinery for DNA synthesis is not entirely preformed in the unfertilized eggs. When the fertilized egg has reached the 4- to 8-cell stage, it has exhausted the pre-existing reserve of DNA precursors (thymidine, in particular). Synthesis of enzymes required for thymidine formation (in particular, ribonucleotide reductase, which transforms ribo- into deoxyribonucleotides) begins at this stage. However, the key enzyme for DNA replication during early cleavage is DNA polymerase α: DNA replication and cleavage are immediately arrested if fertilized sea urchin eggs are treated with *aphidicolin,* a specific inhibitor of this enzyme. DNA replication during cleavage probably follows the same rule as in all other living beings. It is of the semiconservative type (Fig. 45). This means that when the S-period is about to start, the two complementary strands of the DNA double helix unwind and separate. Each strand is copied by DNA polymerase α, so that, in the daughter cells, DNA is made of one pre-existing strand and one newly synthesized strand. Semiconservative replication of DNA has not been proven in cleaving eggs, but there is no reason to think that they might be an exception to the universal rule.

Semiconservative DNA synthesis, during the S-phase, is a *discontinuous* process: it does not go all the way from one end of a chromatid to another in a single continuous step. Instead, there are many *initiation points* (replicons) of DNA replication along the DNA fiber (Fig. 46a), which can be seen by autoradiography

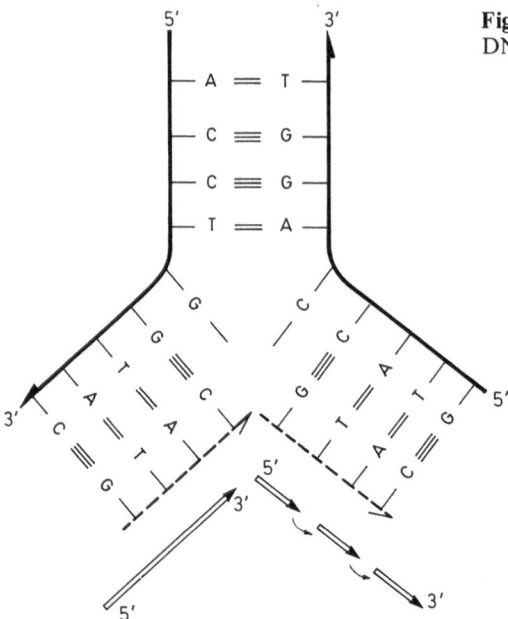

Fig. 45. Semiconservative replication of DNA. (Drawn by Prof. P. Van Gansen)

after pulses of [³H]-thymidine incorporation. At the molecular level the process is more complex than that shown in Fig. 45. As DNA polymerase is only able to polymerize the deoxyribonucleotides in the 5′ to 3′ direction, the DNA replication fork is asymmetrical, i.e. synthesis on the 3′ to 5′ template strand is continuous (*leading-strand template*), while synthesis on the 5′ to 3′ template strand is necessarily discontinuous (*lagging-strand template*). In a first step, a short RNA molecule is synthesized on this separated (now single) DNA strand. This RNA serves as a *primer* for the synthesis of short DNA sequences which are called the Okazaki fragments. These short DNA stretches are then linked together to produce macromolecular DNA. These processes are catalyzed by a not yet completely defined enzyme called *primase,* which synthesizes the Okazaki fragments and participates in linking them together in order to form longer stretches of the DNA fiber. Thus, DNA replication is discontinuous at two different levels: at the chromosomal level, where many initiation points are visible under the light microscope and at the molecular level, by the formation of Okazaki fragments.

If one spreads the chromatin of cleaving eggs and observes it with an electron microscope, one sees numerous "eyes" or "microbubbles" (Fig. 46b), which are regions where DNA is replicating (replicons) and thus the DNA is single-stranded. The microbubbles are clustered in the chromatin of fast cleaving eggs: this means that the main factor responsible for the very fast rate of DNA replication in these eggs is the high frequency of initiation points in chromatin. It does not seem that the speed at which the newly synthesized DNA molecules grow in the replicons is faster in eggs than in somatic cells. One of the reasons for the high

Fig. 46. a Discontinuous DNA replication along a DNA fiber. O and O': Replication origins (initiation points). **b** A "microbubble" seen under the electron microscope. *Arrows* indicate the single-stranded regions in each replication fork (Kriegstein and Hogness 1974)

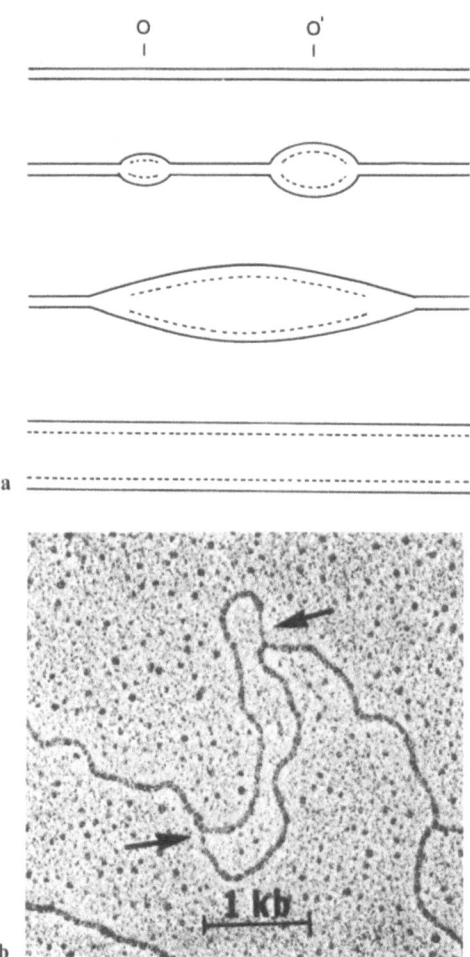

frequency of the initiation points in cleaving eggs is that they possess a DNA synthesis *initiating factor* (IF) which is synthesized during maturation. IF activity decreases when mitotic activity slows down at late cleavage.

We have seen that chromatin is organized in *nucleosomes* which result from interactions between DNA and histones. This leads to the interesting question of *histone synthesis* during cleavage, which has been mainly studied in sea urchin eggs. When DNA is replicated within a few minutes, is chromatin immediately assembled in nucleosomes? If so, where do the histones come from? Unfertilized sea urchin (and amphibian) eggs already possess a large store of "maternal" histones. Nevertheless, fertilization is soon followed by a strong and fast synthesis of all kinds of histones, which is possible because there are between 400 and 1000 histone genes in sea urchins. Histones represent as much as 50% of the proteins which are synthesized during cleavage. Close analysis of the histones which are synthesized during sea urchin egg cleavage and of the corresponding histone mRNAs has disclosed the unexpected fact that there is a *switch* in the varieties of histones which are produced by the developing sea urchin egg. After using first the "maternal" histones and their mRNAs, "early" or "cleavage" histones are synthesized; this is followed, at the blastula stage, by the synthesis of "late" or "embryonic" histones, which are more closely related to the histones found in the adult. Therefore, different sets of histone genes are turned "on" and "off" during the few hours, which separate fertilization from hatching, at the blastula stage. The rate of early histone synthesis increases sharply at the 16-cell stage and reaches a peak at the 128-cell stage; it then decreases gradually until the 300-cell stage where late histone synthesis becomes predominant. In *Xenopus* also, a new set of histone genes begins to operate at the blastula (1000 to 2000 cells) stage.

Recent evidence indicates that, in sea urchin eggs, chromatin is already in the form of nucleosomes at the 2-cell stage: at that time, the nucleosome core is made of early (cleavage) histones which persist in chromatin until the blastula stage. There seems to be a difference in the pattern of histone synthesis between the micromeres (which are important for gastrulation and formation of the skeleton, as we shall see) and the other cells of the 16-cell stage embryo. In addition, chromatin extracted from isolated micromeres is particularly resistant to nuclease digestion, suggesting that it is in a condensed form; it thus presents more similarities with sperm chromatin than with the chromatin of the other blastomeres. It is likely that the shift from "cleavage" to "embryonic" histones during cleavage is responsible for chromatin condensation at the end of cleavage. This would limit the accessibility to the DNA of the enzymes involved in DNA replication and explain the drop in mitotic activity during the blastula-gastrula transition.

We mentioned earlier that amphibian eggs contain a reserve of *cytoplasmic DNA* localized in the yolk platelets. There is no evidence that yolk platelet DNA (in contrast to chromosomal and mitochondrial DNAs) is replicated during development. It is likely that its function is that of a reserve of deoxyribonucleotides which will be made available to the embryo when the yolk platelets break down at later stages of development (gastrulation, neurulation, formation of the young tadpole). Yolk DNA is probably completely broken down into nucleotides when the yolk platelets disappear. However, this is not known for certain and one should mention the fact that if radioactive DNA is injected into a fertilized *Xenopus* egg, it quickly accumulates in the nuclei during cleavage; recovery experiments have shown that part of this DNA is still in an undegraded form in the nuclei. Thus, uptake by the nuclei of large pieces of DNA present in the cytoplasm remains a possibility.

2.4 RNA Synthesis During Cleavage

RNA synthesis is weak, as compared to later stages, during cleavage of amphibian and sea urchin eggs. In fact, there is *no measurable rRNA synthesis* during this period of development in these eggs. They are not making new ribosomes and, thus, they are living on the huge reserve of ribosomes that had been built up during oogenesis. Things are quite different in mammalian eggs, which contain few ribosomes. Synthesis of the rRNAs is easily detectable during early cleavage of mouse and rabbit eggs (see Chap. IX). But, in amphibian and sea urchin eggs, the nucleolar organizers, which are certainly present in one pair of chromosomes, are completely inactive in rRNA synthesis during cleavage. Whether this inhibition is due to the binding of the nucleolar organizers rDNA to a repressor is not known; but this seems a very likely possibility. The absence of measurable rRNA synthesis during cleavage is, of course, in perfect agreement with the fact that true nucleoli are absent during this period of development, except in mammals. In chicken embryos, rRNA synthesis begins a little earlier (midblastula stage) than in sea urchin and amphibian eggs.

In amphibians the synthesis of *mRNAs* starts during early cleavage (8–16-cell stage) and continues, at a low and constant rate, throughout this period. The newly synthesized mRNAs are very heterogeneous in size and represent a whole family of information-carrying molecules. However, at the blastula stage, 85% of overall protein synthesis is still supported by maternal mRNAs. Later, when the embryos synthesize rRNA (at the blastula-gastrula transition) and build up new ribosomes, mRNA synthesis greatly increases. This is particularly striking for the synthesis of the mRNAs coding for the ribosomal proteins. Synthesis of the small RNAs, including 5S RNA and tRNAs, starts at the "midblastula transition" where, after the 12 initial synchronous cleavages, a regular cell cycle makes its appearance. It has been suggested that the egg cytoplasm contains a repressor of RNA synthesis and that this repressor binds to DNA. The cytoplasmic repressor store would be progressively exhausted by the multiplication of the nuclei during cleavage and transcription would start when nuclear DNA is no longer saturated with the hypothetical repressor (Newport and Kirschner 1982). There are, indeed, some indications that cleaving amphibian eggs contain a specific repressor of rRNA synthesis.

In *sea urchin eggs,* the sequence of events is essentially the same, except that mRNA synthesis starts earlier (at about the time of first cleavage) and shows a strong increase already at the 8–16-cell stage. About 60% of the mRNAs synthesized during cleavage code for histones and, as expected, have no (poly A) tail. Curiously, another 30% of the newly synthesized mRNAs (which encode nonhistone proteins) are also devoid of a (poly A) tail: thus, only 10% of the mRNAs synthesized during early cleavage possess the classical (poly A) terminal sequence. At the blastula stage, the situation becomes more normal and 50% of the neosynthesized mRNAs have now the usual (poly A) terminal sequence. Other peculiarities of RNA synthesis in sea urchin eggs are the existence, among the maternal mRNAs, of histone mRNAs which possess a (poly A) tail and the presence in the cytoplasm, during early cleavage, of giant (6.5×10^4 nucleotides) mRNA molecules and of huge polysomes (13.6 µm long). We only know that these giant mRNAs are of maternal origin, since huge polyribosomes can also be found in activated anucleate fragments of sea urchin eggs. Besides, the already discussed histone mRNAs, other specific mRNAs have been identified in the very heterogeneous populations of mRNAs which are synthesized during cleavage. One of them is *actin* mRNA which remains constant during the 8 h which follow fertilization and increases considerably when, after 18 h, the eggs have reached the midblastula (so-called mesenchyme blastula) stage. *Tubulin* mRNA also increases when the blastulae, at the time of hatching, form cilia.

Sea urchin eggs, like all other cells, contain the three different kinds of RNA polymerases which are responsible for RNA synthesis. Polymerase I is specialized in the synthesis of rRNA; its activity increases in parallel with the increase in cell number. Therefore, its activity per cell remains constant. Polymerases II and III, which are more active than polymerase I in the unfertilized eggs, retain a constant activity during the whole cleavage period. This means that their activity progressively decreases on a per cell basis. These two enzymes, like DNA polymerase α are initially present in the cytoplasm of the unfertilized egg and migrate therefrom into the nuclei during cleavage. It is RNA polymerase II which is responsible for the synthesis of new species of mRNAs during cleavage.

A surprise was in store when one examined the effects of inhibitors of RNA synthesis on cleavage. Actinomycin has no effect on cell division either in sea urchin or amphibian eggs. It does not stop cleavage in sea urchin eggs until hatching, i.e., until the late blastula stage, but gastrulation never begins. Still more impressive is the case of the amphibian eggs, in which high concentrations of actinomycin can be injected into the eggs. Without any immediate effect, perfectly normal cell divisions succeed one another repeatedly until the blastula stage is reached; development stops at that stage. It must be concluded that all the information required for repeated cleavage is already present in the unfertilized egg in the form of maternal, masked mRNA molecules. No synthesis of RNA molecules, including those of histone mRNAs, is really needed for repeated mitoses in this case. Microinjection of α-amanitin, which blocks the activity of RNA polymerase II, but not that of RNA polymerase I, also gives interesting results. Cleavage stops at the blastula stage, as after actinomycin injection. Cytological examination has shown that the blocked blastulae have no nucleoli, despite the fact that α-amanitin does not affect the enzyme (RNA polymerase I) which synthesizes the rRNAs and their macromolecular precursors. This experiment shows that synthesis of rRNAs and formation of nucleoli by the nucleolar organizers are not only controlled by RNA polymerase I, but that the derepression of the nucleolar organizer rDNA apparently requires the synthesis of mRNA synthesized by RNA polymerase II. Since, in both sea urchin and amphibian eggs, cleavage occurs, but gastrulation is never initiated in embryos where RNA synthesis has been suppressed, it must be concluded that the RNAs which are synthesized during cleavage are required for later stages of development. The situation is very different in the slowly dividing *mammalian* eggs where synthesis of all kinds of RNAs, including rRNAs, can be detected during early cleavage: addition of actinomycin D quickly arrests their cleavage.

2.5 Protein Synthesis During Cleavage

The level of protein synthesis steadily increases during cleavage, both in the sea urchin and the amphibian egg. Many new proteins are formed, besides the histones. In amphibian eggs protein synthesis follows the animal-vegetal gradient in RNA distribution which was already present in the oocyte. Protein synthesis is stronger at the animal pole than at the vegetal one, which contains fewer polysomes. In sea urchin eggs, it has been shown that both the pre-existing, maternal mRNAs and the newly synthesized RNAs are translated.

The classical inhibitors of protein synthesis, *puromycin* and *cycloheximide* exert the expected effects on cleavage. Mitosis is almost immediately blocked in both sea urchin and amphibian eggs. These inhibitors first inhibit the synthesis of the cytoplasmic proteins; later, DNA synthesis also stops. Injection into amphibian eggs of inhibitors of mitochondrial protein and RNA synthesis (chloramphenicol and rifampicin, respectively) has no effect on cleavage and further development: perfectly normal tadpoles hatch out of the injected eggs. Therefore, mitochondrial DNA can only play a secondary, minor role in early development. On the other hand, it should be evident from what has been said earlier about the effects of anaerobiosis, cyanide and dinitrophenol on cell division that mitochondria play a very important role during cleavage, but as energy producers.

In sea urchin eggs, the first appearance of *new proteins* can be detected at the 16-cell stage, when mRNA synthesis sharply increases. However, many more newly synthesized proteins are found in blastulae. As we have seen, histones are preponderant among them during cleavage. There is little tubulin synthesis before the hatching blastula becomes ciliated: the bulk of the tubulin required for spindle and aster formation at each mitotic cycle thus arises from a pre-existing pool of unpolymerized tubulin molecules. Hatching results from the production of a specific protease, called the *hatching enzyme,* at the blastula stage. In amphibians, the hatching enzyme is synthesized at a much later stage of development (early tadpole). An unanswered question is the following: Is protein synthesis cyclic, in relation to the mitotic cycle? The experiments made on sea urchin eggs have given conflicting results; the conclusion is that, if cyclic variations in protein synthesis exist, they are quantitatively of minor importance only. However, recent work has shown that sea urchin eggs synthesize, at fertilization, a set of three to four new proteins and that one of them, called *cyclin,* disappears and reappears at each cleavage cycle. It might well be identical to MPF, the factor which induces nuclear membrane breakdown and chromosome condensation at maturation and the activity of which also changes in a cyclic manner during egg cleavage.

Among the many factors which control the synthesis of macromolecules during cleavage are probably the *polyamines* (putrescine, spermidine, spermine). These small basic molecules are synthesized at a higher rate whenever cell multiplication increases. They are believed to play a role in the control of RNA and DNA synthesis, perhaps by virtue of their ability to bind to nucleic acids. The key enzyme in polyamine synthesis is *ornithine decarboxylase* (ODC), which converts the amino acid ornithine into putrescine. ODC is one of the few enzymes which are synthesized already during cleavage in both amphibians and sea urchins. It has been reported that there are cyclic changes in ODC activity and in polyamine content during cleavage of sea urchin eggs and that inhibiting ODC activity quickly arrests cleavage in some species of sea urchins, but not in others. The species in which cleavage is quickly stopped by the inhibitors of ODC activity are those in which the fertilized eggs are very poor in polyamines and in which a strong polyamine synthesis takes place already during cleavage. In contrast, the fertilized eggs of the species, in which development is not stopped before gastrulation by the ODC inhibitors, possess a large pool of polyamines and synthesize few polyamines during cleavage. Taken together, these results suggest that polyamines are somehow involved in the control of DNA synthesis at early developmental stages.

2.6 Furrow Formation

In all cleaving eggs, it is clear that the progression of the furrow which will separate the blastomeres requires an *expansion of the cell membrane*. This increase in the surface of the cell membrane does not necessarily imply a synthesis of new membrane proteins. The egg contains a very large store of membranes, present in the form of a discrete endoplasmic reticulum; these membranes might simply be added to the cell membrane when the latter expands during the furrowing process.

In sea urchin eggs, it has been proven that the egg *cortex* has autonomous contractile properties. It has been possible to remove the whole mitotic apparatus, by microdissection, from eggs which were in metaphase or anaphase and it was found that they, nevertheless, divided at the right time. Isolated pieces of the cortex contract if one adds ATP to the medium. These remarkable properties of the sea urchin egg cortex are due to the fact that it contains a *contractile protein*. In fact, all cleaving eggs studied so far possess, in their cortex, bundles of *actin microfilaments*. At telophase, when furrowing begins, these microfilaments form, at the egg equator, a *contractile ring:* its slow contraction progressively divides the egg into two blastomeres. Contraction of the actin microfilaments is inhibited by treatment of the eggs with cytochalasin B: this drug binds to unpolymerized actin molecules and prevents their polymerization. Inhibition of actin polymerization by cytochalasin B is followed by disorganization of the contractile ring of microfilaments. If cytochalasin B is added to an egg in which furrowing is already under way, the furrow immediately stops growing and often regresses. Cytochalasin B has no other effects on mitotic division than inhibition of furrowing, i.e. chromosomes continue to be replicated and mitotic apparatuses are formed in cytochalasin B-treated eggs, resulting in the formation of multinucleated eggs, which ultimately die. While the microfilaments present in the contractile ring certainly play a major role in furrowing, it is likely that other factors are involved in this process. There is some evidence that the tips of the asters at telophase release factors which affect the organization of the cortical microfilaments and induce their assembly in a contractile ring. It is probable that the plasma membrane itself also plays a role, since cleavage of amphibian eggs is inhibited by lectins which bind to carbohydrate residues present on the cell membrane.

Cleavage, in amphibian eggs, presents some particular problems in view of their large size. For instance, a very viscous, gelified area forms around each aster at telophase, whereas the cytoplasm remains fluid and poor in yolk platelets between the two gelified poles. This obviously facilitates the progression of the new membrane by contraction of the cortical microfilament ring. Furrow progression in the bulky amphibian eggs is a slow process which starts at the animal pole and ends at the vegetal one. The nuclei are already in prophase before the furrow has completely divided the egg into two cells. Narcotics impede the periasterial gel formation, which is probably a Ca^{2+} dependent process. As in cytochalasin B-treated eggs, no furrows form, but mitoses go on for several cycles in narcotized amphibian eggs. In addition, delicate experiments on amphibian eggs have shown that if one removes the material, which lies at the tip of the progressing furrow, with a microneedle and injects it into the cortex of another egg near the animal pole, a new furrow forms at the point of injection. Substances present at the tip of the furrow can thus induce a localized contraction of the egg cortex.

In summary, the main purposes of cleavage are the cellularization of the egg and the progressive return of the nucleocytoplasmic ratio to the value it had in the oocyte before maturation. Cellularization is required to provide the egg with the plasticity necessary for morphogenetic movements and inductions which will take place during gastrulation and neurulation. However, gastrulation movements and morphogenetic inductions are still possible in embryos in which, after treatment with various inhibitors of cleavage, the cells are much larger than usual. We shall see in the next chapter that the eggs of the polychaete *Chaetopterus* are even capable of a certain amount of differentiation without cleavage.

Finally, we wish to emphasize again that eggs have played and still play a major role in our understanding of cell division in all cells. For cell biologists, cleaving eggs are ideal materials for the study of DNA replication and its control, for isolation and analysis of the mitotic apparatus, for the induction of nuclear membrane breakdown and chromosome condensation by MPF-like factors (see Chap. IV), for the analysis of furrowing mechanisms, etc. For embryologists, cleavage provides ideal opportunities for the study of the fate of each blastomere. Mere observation allows us to follow the *cell lineage,* in order to find out which part of an embryo derives from a given blastomere. We can destroy at will a given blastomere or culture it in vitro to see whether it is still capable of differentiation when it has been separated from its neighbors. The subdivision between mosaic and regulative eggs (Chap. III) results from experiments done on cleaving eggs. We shall now see what molecular embryology has taught us about germinal localizations in mosaic eggs and regulation in sea urchin eggs.

CHAPTER VII

Molecular Embryology of Invertebrate Eggs

The classical experiments that have led to a distinction between the *mosaic* eggs of worms and mollusks and *regulative* eggs have been briefly described in Chap. III. We know much more about the biochemistry of regulative eggs, because their prototype is that of the sea urchin, which can be obtained easily in very large quantities. It is much more difficult to get enough egg material from most worms and mollusks for biochemical analysis. Therefore, our present knowledge of their molecular embryology remains very scant indeed. Most of the work that has been done on these eggs is based on cytochemical methods or electron microscopy; but these methods cannot give enough information when one wishes to study, for instance, the synthesis of specific kinds of RNAs. A good deal of the work done on invertebrate eggs of the mosaic type has been done by Reverberi and his school. These workers have focused their attention mainly on the *distribution of the mitochondria* during development. Mitochondria are indeed useful markers of certain germinal localizations, where they may accumulate, but there is no evidence that they are the causal agents of embryonic differentiation. In the following, we shall limit ourselves to the few egg species in which biochemical analysis has gone further than mere cytochemical description.

1 *Chaetopterus* Eggs

The eggs of this polychaete are famous among embryologists because they undergo, after activation, what Lillie called *"differentiation without cleavage"*. Treatment of unfertilized eggs with KCl induces maturation, formation of a fertilization membrane and, as usual, building up of a monaster. In activated *Chaetopterus* eggs, several monasterial cycles follow each other: the result is a *pseudocleavage,* where the eggs attempt to divide into blastomeres, but the furrows soon vanish. After these abortive attempts to cleave, the eggs display a strong ameboid activity, which results in a segregation between yolk and clear cytoplasm. In part of the egg population the clear cytoplasm completely surrounds the yolk: this process has been called *overflow* or *pseudogastrulation.* The eggs which have undergone complete overflow of the yolk by the clear cytoplasm are capable of *hatching* and *ciliation,* the result of which is the formation of *unicellular swimming larvae* which digest their yolk reserves and finally become filled with vacuoles. These larvae present a superficial similarity with normal, multicellular larvae (the so-called *trochophores*), however, the unicellular larvae always lack the apical tuft of long cilia and have no intestinal cavity, in contrast to normal trochophores (Fig. 47).

Pseudocleavage and segregation require the integrity of the microtubules; since both result from the repeated formation and disappearance of monasters, the two processes are suppressed by colchicine. Cytochalasin B, in contrast, does not inhibit segregation, however, it prevents ciliation. Treatment with cytochalasin B of fertilized *Chaetopterus* eggs inhibits their first cleavage. The arrested eggs undergo differentiation without cleavage just like the KCl-activated unfertilized eggs.

The successive monasterial cycles lead to repeated chromosome replication and formation of a highly polyploid nucleus. This giant nucleus has the size of the germinal vesicle of a *Chaetopterus* oocyte, but its DNA content is about 300 times higher. However, the rate of DNA synthesis is much lower during differentiation without cleavage than during the development of fertilized eggs. At the time of hatching and ciliation, the large nucleus breaks down into a crown of small nuclei which vary greatly in size and DNA content.

Experiments with aphidicolin, which quickly arrests DNA synthesis in *Chaetopterus* eggs, have shown that inhibition of chromosome replication does not affect pseudocleavage. Formation of a polar lobe and of abortive cleavage furrows is thus an intrinsic property of the egg cytoplasm. Pseudocleavage probably results from an autonomous contractile activity of the egg cortex (which, in *Chaetopterus,* is rich in actin microfilaments and contains 90% of the maternal mRNAs). Since pseudocleavage is synchronous in control and aphidicolin-treated eggs, it must be concluded that the egg cytoplasm possesses a kind of molecular clock which sets the time at which furrowing occurs. If DNA synthesis is continuously suppressed, segregation and the later stages of differentiation without cleavage do not take place. Experiments, in which aphidicolin was added at an early (pseudocleavage) or a late (segregation) stage of differentiation without cleavage, have shown that it is only early DNA synthesis which is important for hatching and ciliation. Addition of aphidicolin to eggs which have been allowed to develop for 5–6 h in normal seawater does not reduce the percentage of swimming larvae, although the DNA content of these larvae is decreased by 50%. Thus, the capacity for the egg to differentiate without cleavage does not depend on the amount of DNA it contains, but on the "quality" of its DNA. A crucial event takes place around the fifth chromosome replication cycle. After this cycle differentiation without cleavage no longer requires DNA synthesis. We shall soon see other cases in which a given DNA replication cycle has a crucial importance for further development. In conclusion, differentiation without cleavage only mimics normal embryonic development since there is no true gastrulation, but it provides an excellent model for the study of cell differentiation, in particular for the analysis of cell motility and cilia formation. When one observes differentiation without cleavage under the microscope, one has – with a little imagination – the illusion that one is following the transformation of an ameba into a ciliate.

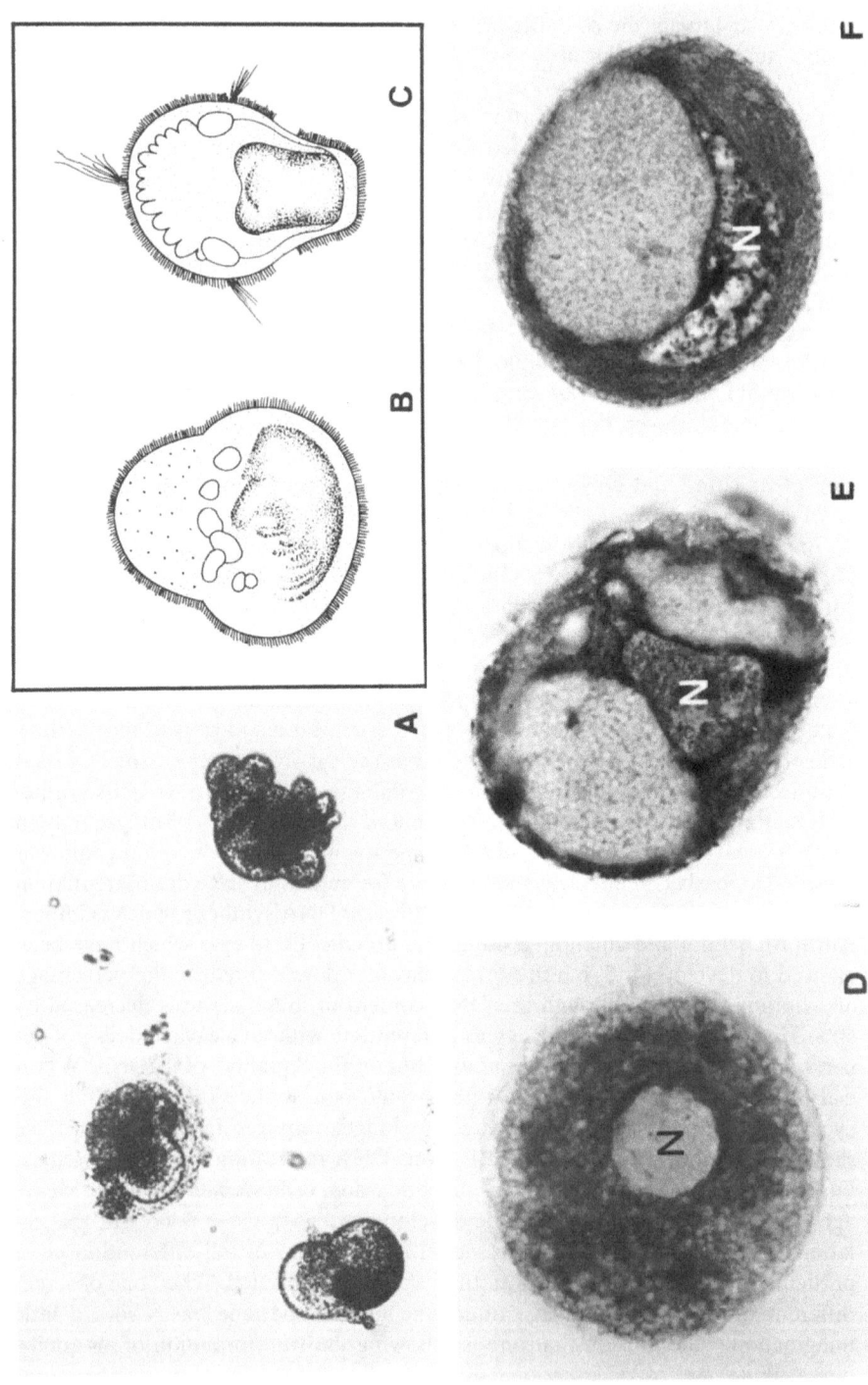

2 Mollusk Eggs

The egg of the gastropod *Ilyanassa* has been better studied, from the molecular viewpoint, than those of other spiral cleavage eggs.

As in many mosaic eggs, mitochondria and lipids accumulate in certain blastomeres during cleavage. On the other hand, the ribosomes and a soluble enzyme, dipeptidase, are evenly distributed despite the complexities of the spiral cleavage.

We have seen that the *polar lobe* in *Ilyanassa* (see Fig. 12) is important for morphogenesis. If it is removed at the trefoil stage, abnormal embryos, with either no foot, shell and eyes or else reduced, are obtained. The polar lobe is rich in yolk platelets and contains many vesicles, surrounded by a double membrane, the significance of which remains obscure.

Polar lobes isolated from *Ilyanassa* cleaving eggs are capable of protein synthesis directed by maternal mRNAs. Removal of the polar lobe affects DNA, RNA and protein synthesis in the remaining embryo, but changes in macromolecule synthesis are late events in such "lobeless" embryos, since they coincide with the appearance of the morphological abnormalities shown in Fig. 12B. Thus, immediate effects of polar lobe removal on the synthesis of macromolecules are not observed. Refined methods of analysis have so far failed to detect polar lobe- specific proteins. Since, in *Ilyanassa,* almost all the proteins synthesized during early embryogenesis are translated from maternal mRNAs, there are no indications so far for a selective localization of translatable, specific mRNAs in the polar lobe.

The pattern of protein synthesis changes, in *Ilyanassa* eggs, between early cleavage and 24-h embryos: new proteins make their appearance, while others are no longer synthesized. Curiously, some of these changes take place in isolated polar lobes, indicating that some of the maternal mRNAs are translated in a selective way: for instance, histones are synthesized at approximately the same rate in intact eggs, lobeless embryos and isolated polar lobes (where there is, of course, no nuclear DNA replication).

These studies, as well as experiments in which the complex effects of actinomycin D on protein synthesis have been studied, lead to the general conclusion that development of *Ilyanassa* eggs, until an early larval stage, does not result from differential gene transcription – as one might have expected – but from translational controls.

Similar results have been obtained in the eggs of other gastropod mollusks, *Lymnaea* and *Dentalium*. In *Lymnaea*, RNA synthesis apparently begins earlier

---

Fig. 47A–F. Differentiation without cleavage in *Chaetopterus* eggs after treatment with a salt solution. **A** Living eggs displaying strong ameboid activity. **B** Differentiation without cleavage at its final stage compared with **C** normal larva from fertilized egg. **D–F** Sections through activated egg before (**D**), during (**E**) and after (**F**) segregation between yolk (*lightly stained*) and basophilic cytoplasm. *N* nucleus (**B, C** redrawn from Lillie 1902)

than in *Ilyanassa;* in fact, during cleavage. In *Dentalium,* the polar lobe which, as in *Ilyanassa,* is very important for morphogenesis, is rich in mitochondria. It also contains DNA-rich inclusions, which are symbiotic bacteria having no special role in development. The presence of symbiotic bacteria in eggs is a warning against hasty interpretation of biochemical results. If, for instance, one found DNA synthesis in an isolated polar lobe, which contains no nucleus, it does not necessarily mean that mitochondrial DNA can replicate in the absence of the nucleus; it could be due to the multiplication of DNA-containing symbionts.

3 Tunicate (Ascidian) Eggs

Ascidian eggs are an ideal material for the study of *germinal localizations,* because they are a mosaic of territories, called *plasms,* which give rise to the different organs of the tadpole larva. Theoretically, two different molecular mechanisms could be imagined to explain how a given germinal localization gives rise to the corresponding organ (how, for instance, the *myoplasm* shown in Fig. 48 gives rise to muscle cells in the tadpole): either maternal mRNAs, coding for muscle-specific proteins are accumulated in the myoplasm; or the myoplasm contains cytoplasmic factors which selectively activate the nuclear genes coding for these muscle-specific proteins, as was suggested by T. H. Morgan in 1934. We would be dealing with preformation in the first hypothesis, and with epigenesis in the second. We shall come back to this important question after we have briefly presented a few facts about ascidian embryology.

The extensive studies of Reverberi and colleagues have shown that there is a close correlation between the formation of muscles in the tadpole larva and the local accumulation of mitochondria. As shown in Fig. 48 the ascidian egg is formed of a number of recognizable "plasms", which give rise to the different parts of the larva, after a precise segregation in the various blastomeres during cleavage. The myoplasm (which will give rise to the muscles) is rich in mitochondria and is accumulated in the two posterior blastomeres at the 4-cell stage. If the egg is centrifuged, soon after fertilization, abnormal larvae, having atypically distributed muscle cells, of which all are rich in mitochondria, are obtained.

Electron microscopy and estimations of cytochrome oxidase activity confirm that the myoplasm-containing posterior blastomeres have three to four times more mitochondria than the anterior cells. However, if one separates the anterior from the posterior blastomeres and measures the respiration of the isolated blastomeres, one finds very little difference. These experiments show that in ascidian eggs, the number of mitochondria present in the individual blastomeres is not the limiting factor for the respiratory rate.

Formation of the different plasms, in ascidian eggs, is a consequence of fertilization. The unfertilized egg is apparently homogeneous and fertilization induces the segregation of the various plasms. For instance, the myoplasm, as well as a large proportion of the mitochondrial population, first migrate and accumulate near the vegetal pole (A); the myoplasm is shifted toward the posterior blasto-

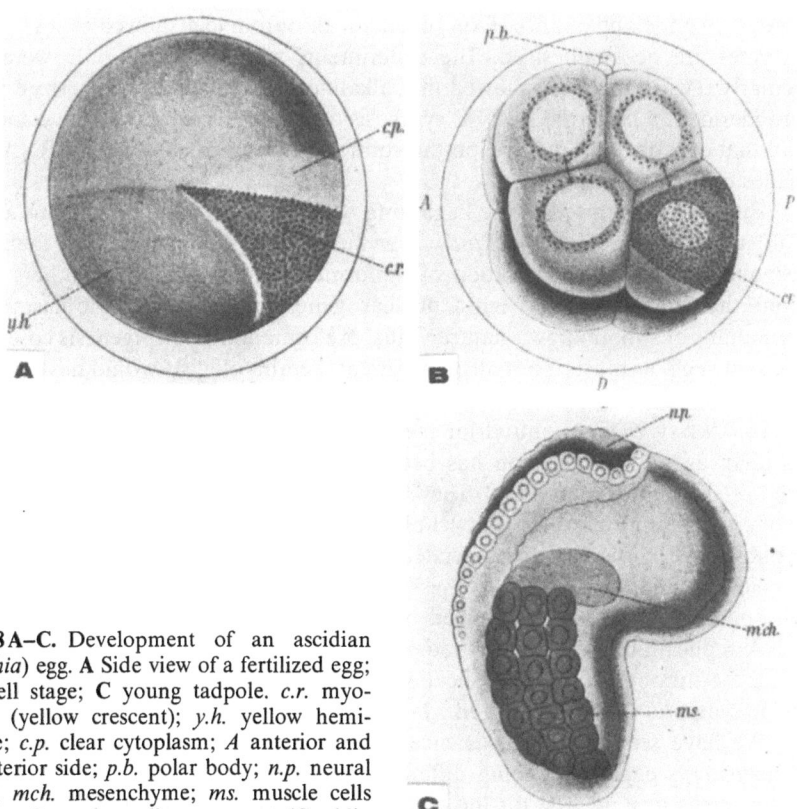

Fig. 48 A–C. Development of an ascidian (*Cynthia*) egg. **A** Side view of a fertilized egg; **B** 8-cell stage; **C** young tadpole. *c.r.* myoplasm (yellow crescent); *y.h.* yellow hemisphere; *c.p.* clear cytoplasm; *A* anterior and *P* posterior side; *p.b.* polar body; *n.p.* neural grove; *mch.* mesenchyme; *ms.* muscle cells deriving from the yellow crescent (Conklin 1905)

meres at the time of first cleavage. Segregation of the plasms, after fertilization, does not take place if the eggs are treated with cytochalasin B, thus showing that the actin microfilament system is involved in plasm segregation. Removal, at the 8-cell stage, of the two posterior vegetal blastomeres which contain most of the myoplasm (B) results in the formation of an abnormal tadpole which lacks muscle cells.

Thanks to the work of Whittaker, marked progress has recently been made concerning the nature of the germinal localizations in ascidian eggs. Using cytochemical methods, he followed the localization of three "marker" enzymes: acetylcholinesterase for muscle, alkaline phosphatase for gut and tyrosinase for the differentiation of pigment cells in the brain. He compared normal embryos, eggs in which cleavage had been arrested by treatment with cytochalasin B and blastomeres isolated at the 8-cell stage. He also used actinomycin D to establish whether the enzymes are synthesized on preformed maternal or on neosynthesized mRNAs. The three enzymes become detectable during cleavage in localized regions of the embryo and at determined times after fertilization. Suppression of furrow formation does not modify the pattern of enzyme localization and the time of enzyme appearance. Experiments with puromycin showed that the three enzymes are neosynthesized. The experiments with actinomycin D were particularly rewarding. They showed that alkaline phosphatase is synthesized in the endoderm on a maternal mRNA; synthesis of the same enzyme, in the ectoderm, and that of tyrosinase require, on the contrary, transcription of new mRNA molecules by nuclear genes.

The problems raised at the beginning of this section have thus found a solution: there are *two different kinds of germinal localizations.* Some of them correspond to a local accumulation of preformed maternal mRNAs; others result from the activation of specific nuclear genes by localized cytoplasmic determinants of still unknown nature. Thus, preformation and epigenesis co-exist in the same egg, as most, if not all, experimental embryologists would have predicted.

Is *DNA synthesis* required for the activation of the germinal localizations in ascidian eggs? This question has been recently studied by Satoh and Ikegami who used aphidicolin as a tool. They found that the eighth DNA replication cycle is of crucial importance for acetylcholinesterase synthesis and concluded that this replication cycle is closely associated with a *clock mechanism* determining the time of initiation of enzyme synthesis. For the other enzymes, including alkaline phosphatase, which is synthesized on a maternal mRNA, a given number of DNA replication cycles is also required before enzyme synthesis begins. However, the number of DNA replication cycles necessary for the initiation of enzyme synthesis varies for each enzyme.

We have seen that the existence of a crucial mitotic cycle is very likely in *Chaetopterus* eggs undergoing differentiation without cleavage. The same situation seems to hold true for the morphogenetic events which take place during the development of mouse eggs. One has to assume that at a given time, a particular type of mitosis (a so-called *quantal mitosis*) occurs and that it gives rise to two nonidentical daughter cells. Unfortunately, we still know nothing about the molecular mechanisms of quantal mitoses which, as we shall see, seem to be important for cell differentiation during embryogenesis.

It has been claimed that development of ascidian eggs until the tadpole stage does not require any RNA synthesis. However, this is very unlikely since actinomycin D inhibits the synthesis of both alkaline phosphatase (in the endoderm) and of tyrosinase. In fact, we have observed that actinomycin D arrests development of fertilized ascidian eggs at the hatching blastula stage. However, we should mention the curious results obtained on *hybrid merogones* between ascidians. In these experiments, eggs of species A were cut into two halves (merogony) and the anucleate half was fertilized with sperm of species B (hybridization). The development of such hybrid merogones, in ascidians, is purely of the maternal, cytoplasmic, A type. In ascidians the sperm nucleus does not exert any specific effect on early development if it is located in cytoplasm from a foreign species. It looks as if the male genes, which are involved in embryonic differentiation, remained silent in the foreign cytoplasm. Unfortunately, hybrid merogones between ascidian species have not been studied so far with biochemical or even cytochemical methods. Therefore, since we know nothing about their RNA and protein synthesis, it is better to postpone a discussion of why development in ascidian hybrid merogones is purely maternal, until more experimental data is available.

4 Insect Eggs

We have already mentioned the interesting *pole plasm* located at the posterior end of *Drosophila* eggs. UV-irradiation of this region results in sterile adults, showing that the cytoplasmic determinants present in the pole plasm (also called *germ plasm*) are absolutely necessary for the formation of the gonads. Injection of pole plasm material from normal eggs into UV-irradiated *Drosophila* eggs allows the formation of fertile adults; injection of the same material into the anterior part of the egg results in the abnormal formation of germ cells in this part. The pole plasm contains large ribonucleoprotein granules, but their role and exact composition remain open to discussion. It has recently been shown that there is a localized synthesis of specific proteins in the *Drosophila* pole plasm: this suggests that either a differential gene process or selective translation of maternal mRNAs takes place in this region.

Classical experiments by Seidel, in which insect eggs were cut into two parts by ligation with a hair, have shown that development, in several species, requires an interaction between one of the cleavage nuclei and cytoplasmic determinants located in the posterior half of the egg. After penetration of a nucleus, whether it was originally present in the anterior or posterior part of the egg, the posterior region becomes a *formation center (Bildungszentrum):* it produces diffusible substances which are necessary for the development of the anterior part of the egg. Unfortunately, nothing is known about the chemical nature of the cytoplasmic determinants present in the formation center or about the diffusible substances it produces after it has been activated by a nucleus.

A little more is known about the eggs of the midge *Smittia,* in which development of the head also depends on cytoplasmic determinants located in the posterior half of the egg. If one UV-irradiates or injects this posterior half with ribonuclease, one obtains "double-abdomen" larvae which possess abdominal structures instead of a head (Fig. 49). This suggests that the cytoplasmic determinants of *Smittia* are of a ribonucleoprotein nature and that maternal mRNAs localized at the posterior end of the egg are involved in the differentiation of the anterior part. It has recently been shown that *Smittia* eggs synthesize a specific "posterior indicating" protein. However, the presently available evidence is still too scanty to allow strong conclusions about the chemical identity of insect cytoplasmic determinants and their mode of action.

Although we are dealing in this book mainly with early stages of development, mention should be made on the important work done on *Drosophila* (an ideal material for geneticists and cytogeneticists) larvae. The future appendages of the adult (antennae, eyes, etc.) are, in larvae, present in the so-called *imaginal*

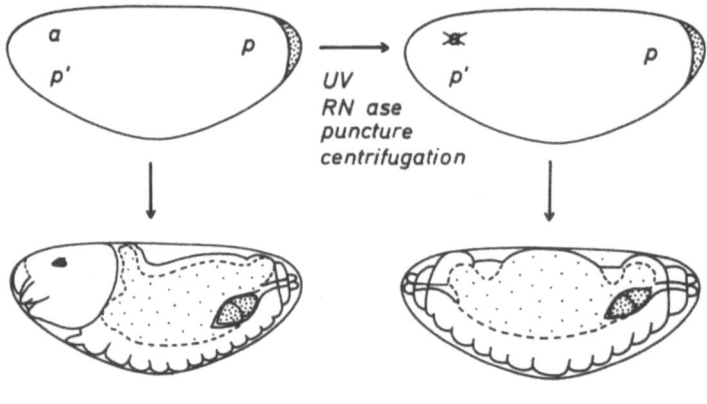

Fig. 49. *Double abdomen* induction in *Smittia* under various experimental conditions which inactivate anterior determinants (*a*) *p, p'* posterior determinants (Kalthoff 1979)

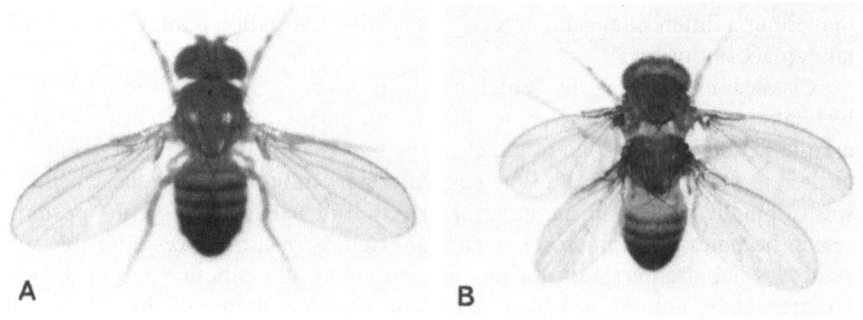

Fig. 50 A, B. Normal (**A**) and bithorax (**B**) fruitfly (*Drosophila*). (Courtesy of Dr. E. Lewis)

disks made of a limited number of undifferentiated cells, e.g. antenna, eye, etc. disks which will normally differentiate into, respectively, antennae, eyes, etc. However, transplantation experiments by E. Hadorn have shown that it is possible to modify experimentally the fate of the imaginal disks, i.e. an eye disk may, under certain circumstances, differentiate into an antenna or leg. To this phenomenon, Hadorn has given the name *"transdetermination"*. A similar change from one type of appendage to another can result from a genetic mutation. Such mutations, which transform one segment of the insect into another, are called *"homeotic"* mutations. One interesting example of a homeotic gene mutation affects the *bithorax* complex located on chromosome III. The extreme mutant bx^3 has four wings instead of two (Fig. 50): in the segment bearing them, the halters have been transformed by the mutation into wings. Analysis at the molecular level has recently shown that the bithorax mutations are due to rearrangements of the DNA molecules and not, as it was believed, to point mutations (substitution of a single nucleic acid base by another). Many bithorax mutants display insertions of a particular mobile element (called *gypsy*) into their DNA; such insertions affect the functioning of DNA sequences distant from them.

Several kinds of transposable elements (virus-like DNA sequences which may be inserted anywhere in the genome as we have seen in Chap. II) have been identified in *Drosophila*. Their function remains a matter of speculation, but it is likely that they will receive more and more attention from development geneticists. Recent work has shown that injection of a transposable element called *P* into the germ plasm of *Drosophila* eggs induces heritable mutations in strains which lack *P* elements. If one inserts a *Drosophila* eye-color gene into a transposable *P* element, this gene can be introduced into recipient eggs: thus transposable elements can be used as *vectors* for gene transfer in *Drosophila*.

It has also been reported that insertion of a *P* element into the RNA polymerase II gene is lethal, i.e. the embryos die at an early stage because the enzyme, which is needed for mRNA production, is no longer synthesized.

It has been recently discovered that six genes, which control the differentiation or the number of segments in *Drosophila* larvae, share a common short (180 base pairs) DNA sequence. It encodes a very basic polypeptide made of 60 amino acids, which binds strongly to DNA. This sequence, which has been called the *homeo box,* probably controls the subdivision of the larval body into segments and the diversification of these segments. One of the genes, which possesses a homeo box (ft3), is transcribed already during cleavage. At this stage, its RNA transcripts are restricted to seven evenly spaced bands of cells, corresponding to the future larval segments; the transcripts disappear at later stages. Unexpectedly, the same *homeo* sequence is found in the nuclear DNA of many invertebrates and vertebrates (*Xenopus,* chicken, mouse, man). It is not found in species in which the embryo does not undergo segmentation (metameric organization) during development (sea urchins, for instance).

There is little doubt that a *homeo box* is required for the control of the subdivision of the body into segments in all animals.

After this short incursion in the fascinating field of developmental genetics, we shall return to our main topic and have a look at sea urchin molecular embryology.

5 Sea Urchins

Before going into experimental and molecular embryology of sea urchin eggs, a brief description of their gastrulation should be given. As shown in Fig. 51, the blastulae, after hatching and ciliation, become *"mesenchyme blastulae"* characterized by the presence of a thickened *basal plate* which derives from the micromeres present at the vegetal pole at the 16-cell stage. Some of the basal plate cells move into the blastocoele cavity where they form the *primary mesen-*

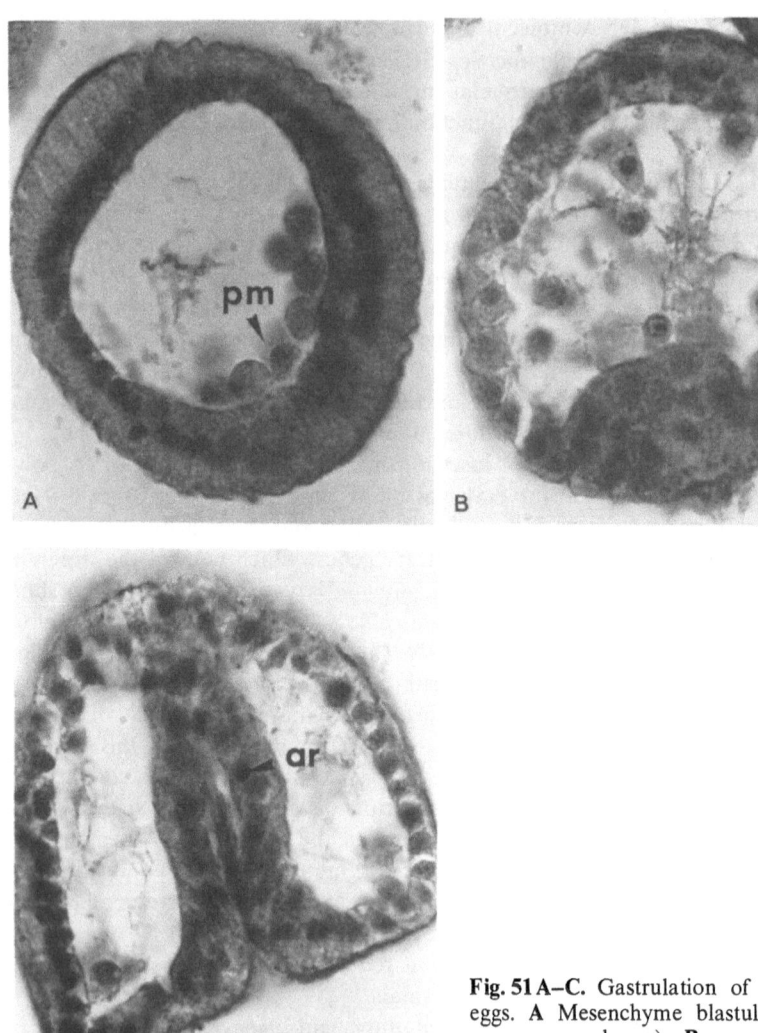

Fig. 51 A–C. Gastrulation of sea urchin eggs. **A** Mesenchyme blastula (*pm* primary mesenchyme); **B** very early gastrula; **C** late gastrula (*ar* archenteron)

chyme, which will give rise to the skeleton of the *pluteus* larva. This skeleton is made of *spicules,* which have a species-specific shape. Growth of the spicules is responsible for the shape of the plutei. Cultured micromeres can give rise to spicules in vitro. Gastrulation begins by an infolding (*invagination*) of the basal plate, which forms the *archenteron* (primitive gut). After this "primary" gastrulation, cells at the tip of the archenteron move into the blastocoele where they form the *secondary mesenchyme.* These cells link together the tip of the archenteron to the apical ectoderm; their contraction produces the complete invagination of the archenteron.

Many biochemical changes occur during sea urchin gastrulation, which requires both RNA and protein synthesis as shown by the inhibitory effects of actinomycin D and puromycin. In particular, the pattern of protein synthesis changes considerably and many proteins are glycosylated and sulfated during gastrulation. Lack of sulfate ions in the medium prevents gastrulation, which is also inhibited by drugs (tunicamycin, compactin), which block protein glycosylation. Actin mRNA, as we have seen, increases in mesenchyme blastulae; this increase is probably related to cell contractility and motility, which are prominent features of gastrulation. DNA synthesis, as shown by experiments with aphidicolin, is required only for preparation to gastrulation. During gastrulation itself, there are very few mitoses and an increase in cell number is not necessary for the invagination of the archenteron.

Sea urchin blastulae are easily dissociated into individual cells by culture in Ca^{2+} free seawater. After adding back Ca^{2+}, the cells reaggregate and form larvae which look like normal gastrulae and plutei. If one mixes dissociated cells from blastulae belonging to two different species, they first aggregate and then separate (*sorting out*) according to their species specificity. The supernatant of dissociated sea urchin blastulae contains a species- and stage-specific *aggregation factor*, i.e. a lectin, which binds the cells together by reacting with carbohydrate residues present on the cell surface.

In Chap. III we already mentioned that sea urchin eggs display, at the 2-blastomere stage, an exceptional capacity for regulation. As shown in Fig. 15, separation of the two first blastomeres is followed by complete and normal development, until the pluteus larval stage, of each blastomere (Driesch). On the other hand, equatorial sections at the morula stage have very different consequences. The development of the animal and vegetal isolated halves, as shown in Fig. 16, is entirely different (Hörstadius). *Animalization* (that is, a development similar to that of the animal half) can be induced in whole eggs by a great variety of agents; *vegetalization*, which represents the opposite type of development, is much more difficult to obtain by treatment of whole eggs with chemichals. Only *lithium chloride* (LiCl) gives really satisfactory vegetalization, identical with that obtained by the much more delicate method of cutting the egg into two halves (Fig. 16). As one can easily imagine, molecular embryologists, who require a large amount of material for their experiments, have done little work on isolated halves of morulae, they have preferred to work on a large scale, treating whole eggs with animalizing or vegetalizing agents. The very careful and painstaking work of Hörstadius, who first succeeded in cutting sea urchin eggs into animal and vegetal halves, led him to the conclusion that the sea urchin egg contains two opposite, almost hostile, *gradients*. The animal gradient decreases from the animal toward the vegetal pole; the vegetal gradient decreases in the opposite direction. Normal development, until the pluteus stage, can only be obtained if the balance between the two opposite gradients remains correct. Especially important for morphogenesis, as Hörstadius demonstrated, are the *micromeres,* which are the most vegetal of all the blastomeres. Grafting enough micromeres to an isolated animal half will completely correct the animalization and the pluteus will be normal (Fig. 16); micromeres can induce gastrulation and the formation of an archenteron at the location of the animal half where they have been grafted.

This very brief account of sea urchin descriptive and experimental embryology was necessary before we could discuss its molecular embryology. Two main questions should be asked: What is the chemical nature of the animal and vegetal gradients? What is the molecular basis of embryonic regulation? These two questions will be discussed consecutively.

The question of the *chemical nature of the morphogenetic gradients* present in sea urchin eggs has been extensively studied by Runnström and his Swedish colleagues. Lindahl first studied the effect of animalizing and vegetalizing agents on the *oxygen consumption* of sea urchin eggs. He found that LiCl has no effect on the respiratory rate of the just fertilized egg, but it slows down the increase in respiration which occurs during cleavage. From these and other experiments, Lindahl suggested the following hypothesis: Metabolism would be qualitatively different at the two opposite poles of the egg; carbohydrate metabolism would be prevalent at the animal pole and protein metabolism at the vegetal one.

The experiments made to test the validity of this hypothesis have given conflicting results, so that no strong conclusion can yet be drawn. For instance, it was found that the oxygen consumption of isolated animal and vegetal halves is the same and that lithium does not exert, as one would have expected, a stronger inhibitory effect on the respiration of the animal than the vegetal halves. On the other hand, Hörstadius has shown the existence of two opposite *reduction gradients* in sea urchin blastulae. If an egg is vitally stained by a dye, which becomes colorless when it is reduced, and if this egg is placed under anaerobic conditions, two centers of reducing activity can be seen; they are localized at the animal and the vegetal pole. An isolated animal half has only one reducing center, located at the animal pole; an isolated vegetal half also has only one reducing center, but it is localized at the vegetal pole this time. If micromeres are grafted on an isolated animal half, a new reducing center develops around them. The biochemical significance of such reducing centers is unfortunately obscure and complex.

The experiments designed to prove that the vegetal pole is a region of high protein metabolism have also given contradictory results. It has been reported that incorporation of amino acids into proteins, as judged by autoradiography, is higher at the vegetal pole (in particular, in the micromeres) than in the rest of the embryo. However, biochemical analysis has failed to confirm the autoradiographic results. Protein synthesis increases during cleavage, levels off when cilia form at the time the blastulae hatch and increases again. But, according to Berg, there is no difference between isolated animal and vegetal halves as far as protein synthesis is concerned. Furthermore, isolated micromeres do not synthesize more proteins, per volume unit, than the other blastomeres. Berg has also studied the effects of LiCl on protein synthesis. It does not inhibit the synthesis of proteins before the end of cleavage, but then the larvae become abnormal; no differential effect of LiCl on protein synthesis in isolated animal and vegetal halves could be detected. More recently, modern methods for the separation of neosynthesized proteins have been applied to blastomeres isolated at the 16-cell stage: the same proteins are synthesized by micromeres, mesomeres and macromeres. Thus, no differences in gene expression between micromeres and the other blastomeres can be detected with the now available techniques. However, molecular differences between micromeres and the other blastomeres certainly exist. We have already seen that the pattern of histone synthesis is quantitatively (but not qualitatively) different in micromeres as compared to other cells and that the chromatin of micromeres is more resistant to nuclease digestion than that isolated from the other blastomeres. This suggests that, in contrast to what one might have thought, gene expression is decreased in the micromeres as compared to the other blastomeres. It has also been shown that the mRNA population (mainly maternal mRNAs) is less complex in micromeres than in the other blastomeres. Finally, it has been reported that in embryos dissociated at the 16-cell stage, the micromeres undergo selective aggregation, i.e. the chemical composition of their cell surface thus differs, in some unknown respect, from that of the other blastomeres. One must admit that, on the whole, the attempts made to discover the molecular bases for the important role played by the micromeres in sea urchin embryogenesis have so far been frustrating; but this is not a reason for despair and the problem will certainly receive further study.

For later stages of development (gastrulation, formation of the pluteus) biochemical analysis has been more successful. It has been possible to show that the ectoderm and the endoderm both synthesize a small set of tissue-specific proteins. Since the mRNAs, corresponding to these proteins, are distributed in the same tissue-specific manner, it can be concluded that ectoderm- and endoderm-specific genes are activated during sea urchin embryogenesis.

A few years ago, attempts to isolate animalizing and vegetalizing factors from whole unfertilized sea urchin eggs were made. It was reported that it is possible to separate, by chromatography, peptides which have either animalizing or vegetalizing activities. Unfortunately, research in this field was discontinued after the death of J. Runnström and the retirement of S. Hörstadius. The animalizing and vegetalizing peptides have remained poorly characterized and we do not even know whether the animalizing peptides are accumulated, at the morula stage, in the animal half and the vegetalizing ones at the vegetal pole.

The second basic question we asked about sea urchin molecular embryology had to do with the classical experiment of Driesch on embryonic regulation. If the two first blastomeres can form a whole embryo when they are separated and only a half embryo when they are kept together, it seems logical to assume that the material which holds the two blastomeres together must have inhibitory properties on development. We have tested this possibility experimentally. The intercellular cement, which keeps the blastomeres together, can be removed from the eggs by placing them in calcium-free and magnesium-free seawater. Sea urchin morulae have been placed in such a medium in order to dissociate the blastomeres. The "conditioned medium" in which the dissociated blastomeres had developed for 1 to 2 h was recovered and calcium and magnesium were added back. The effects of such conditioned media were tested on "recipient", just fertilized eggs. It was found that they arrest development of the recipient eggs at the *late morula or blastula* stage. The experiments prove that the intercellular matrix, which holds the blastomeres together has, as we had expected, *inhibitory* effects on sea urchin egg development. Inhibitory conditioned media could be prepared from unfertilized and fertilized eggs, as well as blastulae; but conditioned media prepared from dissociated gastrulae had no effect on the development of normally fertilized, recipient eggs. This shows that the matrix, which links the cells together, undergoes quantitative or qualitative changes during development. Conditioned media do not display any strong species-specificity. A medium prepared from *Paracentrotus* eggs will inhibit the development of fertilized *Arbacia* eggs, but in a slightly less effective way. Heating the conditioned media at 60 °C reduces, but does not destroy, the inhibitory activity. It is noteworthy that conditioned media prepared from morulae often exert, even after heating, a strong *animalizing* effect. This finding suggests that among the constituents of the intercellular matrix, which is mostly made of glycoproteins, thermostable animalizing polypeptides, which have been found in unfertilized eggs, might be present. It is too early to say whether continuation of this work will lead to a real understanding of embryonic regulation is sea urchin eggs. But, however incomplete the present results might be, they provide possibilities for further research, which could not be found in Driesch's philosophical concept of the "entelechy".

CHAPTER VIII

Molecular Embryology of Amphibian Eggs

This chapter will deal almost exclusively with *amphibian* eggs, since relatively little work has been done on the eggs of other groups of vertebrates. However, important studies have been made on the eggs of a fish, the loach *Misgurnus fossilis* by Russian workers, and have been reviewed by Kafiani. Very little is known about the reptiles in the field of molecular biology. Considerable work has been done on chicken embryos, but at late stages of development. The partial cleavage and the small size of the embryos, as compared to the huge mass of the yolk, makes the material a difficult one for molecular embryologists interested in early development. The mammalian embryos, as already mentioned, must be cultivated in vitro, so that only very limited amounts of material are available for biochemical studies, which will be discussed in the next chapter.

The main periods of amphibian development that will be studied here are those of *gastrulation,* characterized by intensive cell movements, and *neurulation* where, as the result of an induction, primary organogenesis occurs. The neurula contains a closed nervous system, a chorda (notochord), mesoderm which subdivides into somites (which will form principally the muscles and the skeleton), the intermediary pieces (which will give rise to the kidneys) and the lateral plates (which will form the red blood cells and the peritoneal cavity of the adult). The mesoderm, in the early neurula is still undifferentiated. Figure 52 represents sections through a neurula; Fig. 14 has already summarized gastrulation. For more

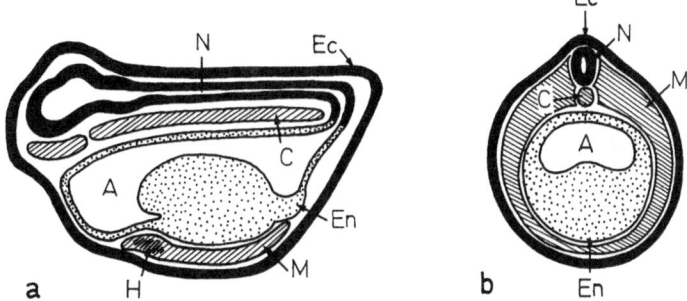

Fig. 52 a, b. Amphibian neurulae. **a** Longitudinal section; **b** transverse section. *A* archenteron; *C* chordoblast; *Ec* ectoderm; *En* endoderm; *H* heart-forming cells; *M* mesoderm; *N* nervous system

details, the interested reader should consult one of the many textbooks of descriptive and experimental embryology.

The main problem that will be discussed is that of the *molecular aspects of neural induction.* Its experimental basis has already been presented in Chap. III in which one of the most important experiments for molecular embryology has been shown in Fig. 17. If a gastrula ectoblast is explanted and cultivated in vitro, it differentiates into ectoderm (skin) only. If the same piece of ectoblast is put in contact with the organizer (dorsal lip of the blastoporus, which will differentiate later into chorda), it is transformed, by induction, into a nervous system. But, before coming to the main problem, it is necessary to mention the *respiration* of the amphibian egg during the gastrulation and neurulation periods.

1 Respiration of the Amphibian Gastrula and Neurula

The oxygen consumption, which had slowly increased during cleavage, shows a much sharper increase during gastrulation and, especially, neurulation. The effects of anaerobiosis and cyanide become stronger than they were during the cleavage period. Development progresses for a very short time only if the energy produced by the cellular oxidations is suppressed.

The *respiratory quotient* (CO_2/O_2) increases at the time of gastrulation. This shift suggests that the gastrula uses mainly carbohydrates as an energy source. Indeed, it was found that glycogen is not broken down to a measurable extent during cleavage, and that glycogenolysis becomes important during gastrulation and neurulation.

Painstaking work has been done in order to measure the respiration of *fragments* isolated from different parts of the gastrula. Eggs have been cut, as shown in Fig. 53, into six different pieces, which correspond respectively to (1) dorsal ectoblast (also called neuroblast, for it will normally give rise to the nervous system); (2) ventral ectoblast, which will form the skin; (3) chordoblast, the "organizer"; (4) ventral mesoblast, which has no inducing capacity; (5) dorsal entoblast; and (6) ventral entoblast. The oxygen consumption of these six pieces de-

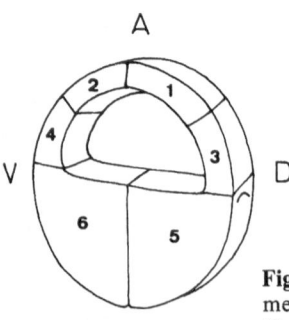

Fig. 53. Dissection of a young amphibian gastrula before measurement of the oxygen consumption of the different parts (Sze 1953). *A* animal pole; *D* dorsal side; *V* ventral side; *Vg* vegetal pole

creases according to a very simple rule: $1 > 2 > 3 > 4 > 5 > 6$. This means that there is an animal-vegetal gradient as well as a dorsoventral one, in the respiratory rate. The dorsal lip of the blastopore (piece n°3) differs from the ventral mesoblast (piece n°4), which will invaginate a few hours later and form the ventral lip of the blastopore. It is the site of a more active carbohydrate metabolism.

It was at first thought that the high carbohydrate metabolism of the dorsal lip might have something to do with its inducing activity. This idea is no longer accepted and it is now believed that the high carbohydrate metabolism of the dorsal lip is required for the achievement of the *morphogenetic movements*. As we already know, invagination begins earlier in the dorsal than in the ventral lip and this cell movement is accompanied by a breakdown of the glycogen reserves in both lips of the blastopore.

Induction of the nervous system is, however, accompanied by an increase in oxygen consumption. If, as we have seen, pieces of ectoblast and chordoblast are placed in close contact, neural induction will follow; if the same pieces are not in contact, but are kept apart, there will, of course, be no induction. The respiration of such pieces is higher and increases more steeply in the first type of combination (contact between the fragments) than in the second (separated fragments). This is not surprising, since, as well shall see, induction is characterized by a burst in nucleic acid and protein synthesis; these processes require the intervention of ATP, the synthesis of which is coupled with oxidations in the mitochondria. The latter enlarge, become structurally more complex and have a higher cytochrome oxidase content during neural induction.

The fact that energy production is required for morphogenesis is also very clear in bird and mammal eggs. Very young chicken embryos can be removed from the yolk and cultivated in vitro, where the nervous system, chorda, somites, etc., will differentiate; but the culture is unsuccessful unless glucose or another metabolizable sugar is added to the simple salt-containing medium. Intermediate products of carbohydrate metabolism must also be added to explanted mammalian embryos in order to obtain cleavage and further development.

2 The Nature of the Inducing Substance

Ever since it was discovered, around 1930, that a boiled or alcohol-treated dorsal lip of the blastopore still remained capable of induction (although less normal than with the living tissue), attempts have been made to isolate the active inducing substance. This search has long given frustrating and disappointing results for unexpected reasons. The ectoblast cells react "too well" and undergo neural transformation, even when they are treated with only unspecific stimulating agents.

The first experiments, mostly by Holtfreter, showed that almost any plant or animal tissue, if it is placed in close contact with "competent" cells (that is, cells taken from the animal pole region of a young gastrula), can induce nervous system formation. Only such very inactive tissues as banana peel or insect wings failed to elicit a reaction! It was concluded that the inducing substance (also called *evocator*) is present in all tissues and that it should be a relatively easy matter to isolate and purify it from large organs such as beef or horse liver.

It soon turned out that many unrelated substances, such as sterols, fatty acids, glycogen, nucleotides and ATP, were all active in inducing the formation of a nervous system. It was first thought that such unspecific results were due to the presence of some common contaminant in the preparations used for the biological tests. But it turned out, around 1935, that even purely synthetic compounds, such as the basic dyes, methylene blue and neutral red, could also be active. This finding led to the conclusion that all these chemicals act in an *indirect* way. They would release in an unspecific way, the true inducing agent, which would be present in the competent ectoblast cells in a *masked,* complex form.

Most chemical embryologists, who were interested in the problem, gave up the search when Holtfreter reported that a simple shift in the pH of the salt solution in which the pieces of ectoderm are cultivated is enough to produce numerous cases of spontaneous transformation of the ectoblast into nerve cells. No less distressing was the fact that simple salts, such as LiCl or NaCl, crushed glass or Teflon, are also good inducers of nervous system formation. In fact, even CO_2 and ammonia are active, so that one had to come to the sad conclusion that the organizer might act simply by producing these simple end products of its metabolism. They would act by the above mentioned "release" mechanism and liberate an unknown "neurogenic" substance from an inactive complex present in the cells of the competent ectoblast.

It took courage on the part of people like Yamada, Toivonen and Tiedemann to come back to the problem, around 1950. They selected amphibian species in which spontaneous neuralization of explanted ectoblast is difficult to obtain and in which the simple blowing of CO_2 into the medium is insufficient to transform ectoblast into nerve cells. They looked for *proteins* as the specific inducing agents and they could at least prove one point. It is possible to isolate, in almost pure form, proteins that have *different inducing specificities.* A soluble protein, which can be extracted and purified from bone marrow and from chicken embryos, acts as a vegetalizing factor. It transforms a gastrula ectoblast into mesodermal tissues, such as chorda, or muscle and endoderm derivatives. The ectoblast, which would normally have formed epidermis (or nerve cells, if in the presence of an inducer), becomes a tail, which is devoid – or almost entirely devoid – of a nervous system. Another protein, which is associated with the ribosomes of the chick embryos, is a *neuralizing* factor. At low concentrations, it transforms the ectoblast cells into neural cells with a high efficiency. A third protein is an *inhibitor* of the vegetalizing factor and might control its activity.

A vegetalizing factor, isolated from chick embryos by Tiedemann and coworkers, has been purified to a considerable extent. This factor (a glycoprotein) is a valuable tool for the analysis of the transformation of ectoderm into mesoderm or endoderm. Continuous culture of an ectodermal explant (taken at the early gastrula stage) in the presence of the vegetalizing factor transforms it into endoderm and suppresses the formation of cilia. Short treatments of the ectoderm with the vegetalizing factor induce the formation of red blood cells and pronephros (embryonic kidney); longer treatments lead to the formation of pronephros, somites and chorda. If somites and chorda are present in large amounts, a nervous system may be induced because chorda produces a neuralizing factor.

One can hardly doubt that although the danger of a release mechanism can never be completely excluded, different proteins have distinct effects on morphogenesis and that these effects are *specific*. This finding makes it very likely that in the normal amphibian embryo itself, different proteins are responsible for the differentiation of the nervous system and of mesodermal tissues. In fact, it has been reported that amphibian embryos indeed contain a vegetalizing factor which is biologically similar to that isolated from chick embryos. It would be important to know, in order to complete the demonstration, whether the vegetalizing factors isolated from bone marrow, chick embryos and amphibian gastrulae are *chemically* similar and belong to a same class of proteins.

As one can see, there are close analogies between two of the problems that have just been discussed: animal and vegetal gradients in sea urchin eggs and neural induction in amphibian eggs. In both cases, the biological system is exceedingly sensitive to the effects of such simple agents as ions or small organic molecules; this excessive sensitivity has been and still remains a very serious obstacle for molecular embryology. However, the attempts made in recent years to isolate *specific* biologically active substances from the eggs themselves have already been rewarding. Their pursuit should make possible an analysis of the molecular bases of embryonic differentiation in both sea urchin and amphibian eggs.

3 RNA Localization and Synthesis

We already know that in amphibians, the ribosomes are distributed along a decreasing *animal-vegetal* gradient in the oocyte, the unfertilized egg and the blastula; part of these ribosomes are present in the form of polysomes, since protein synthesis follows the same gradient and progressively decreases from the animal to the vegetal pole. The distribution of the mRNAs in fertilized amphibian eggs is not yet known with certainty, since there are contradictory reports in the literature. What is clear is that this distribution is not homogeneous in oocytes and eggs and that it follows a polarity gradient. This preformed polarity gradient is important for future development, since the head of the tadpole and the adult will form at the animal pole and the belly at the vegetal pole. It corresponds to the *antero-posterior, cephalocaudal* gradient of the embryo.

Dorsoventral organization is much less evident during these early stages of development. We have seen that in certain species, a *grey crescent* appears in the fertilized egg. The localization of this grey crescent, which roughly corresponds to the future dorsal lip of the blastopore (thus to the chordoblast, which will later induce the formation of the nervous system), can be established at will by the experimenter (Fig. 54). If the unfertilized egg is placed in an oblique position and

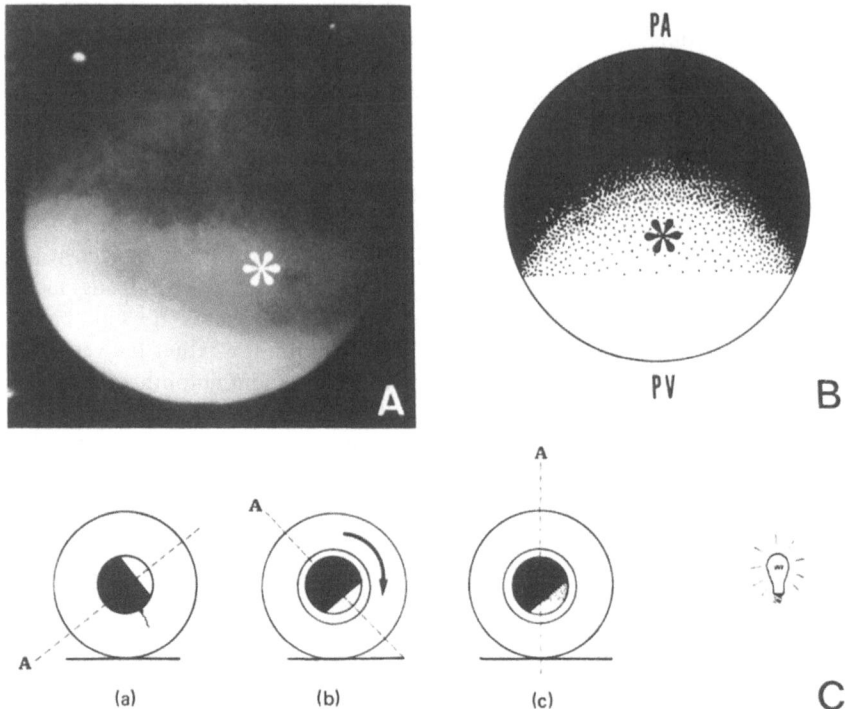

Fig. 54 A–C. The grey crescent of amphibian eggs. **A** Photograph and **B** schematic representation of a grey crescent (*asterisk*). **C** The classic experiment of Ancel and Vintemberger (1948): **a** An unfertilized frog egg is placed in an oblique position and is then fertilized. *A* animal pole. **b** The egg rotates under the influence of gravity after the elevation of the fertilization membrane, as indicated by the *arrow*. **c** The grey crescent always appears on the side facing the lamp. (**A** courtesy of Dr. M. Namenwirth)

fertilized, it will rotate under the influence of gravity after the elevation of the fertilization membrane, as indicated by the arrow. After the rotation, the animal pole will take an upward position, the vegetal pole will place itself downward; the grey crescent, and later the dorsal lip of the blastopore, will always appear on the side marked on Fig. 54 c by the lamp. Of course, this is not due to a photodynamic effect and the results would remain the same whether the electric light is on or off. The formation of they grey crescent is due to *cortical* changes, which occur during the rotation of the egg.

The grey crescent results from a localized thinning out of the pigment layer, due to a displacement of the yolk mass by gravity or growth of the spermaster: the yolk forms a "vitelline wall" which locally compresses the egg cortex. It would be interesting to know whether fertilized amphibian eggs placed in a Spacelab (where there is almost no gravity) would form a grey crescent and whether normal development would be possible.

UV-irradiation of the grey crescent results in the formation of abnormal tadpoles, in which the nervous system (in particular, the brain) is very poorly represented: *microcephaly* is the rule in such experiments, suggesting that material localized in the grey crescent is important for neural induction. UV-sensitivity of this material suggests, but does not prove, that nucleic acids or nucleoproteins are involved in nervous system induction. However, the grey crescent is only a marker of dorsoventrality: if one keeps fertilized eggs upside down (thus with the vegetal pole upwards), a second grey crescent appears on the ventral side; double embryos result from such inversion experiments.

Precocious formation of a grey crescent can be obtained in the axolotl: if oocytes undergoing maturation are injected with inhibitors of protein synthesis (diphteria toxin or cycloheximide), they form a grey crescent. This does not occur if the oocytes were enucleated prior to injection of the inhibitor; but a grey crescent is formed, in such anucleate oocytes, if sap from the germinal vesicle is injected either before or after diphteria toxin treatment. This suggests that the nuclear sap of the germinal vesicle, which is released in the cytoplasm at maturation, plays a role in grey crescent formation and perhaps in later events, which ultimately lead to nervous system induction.

If, as already mentioned, the cortex of the grey crescent is injured by pricking, very deficient embryos with a poorly developed chorda and nervous system will form. It would seem logical to assume that the dorsal lip of the blastopore is more active in RNA and protein synthesis than the more ventral regions of the gastrula, since it is so active for morphogenesis because of its inducing activities. If such were the case, the grey crescent might be endowed with special properties even in the fertilized egg, enabling it to exert some kind of control on the activity of the nuclei in the dorsal half of the embryo.

Unfortunately, experiments made in order to test this hypothesis have not given the clear-cut results one had hoped, for an unexpected reason. Slight injury of the grey crescent and, to a lesser extent, of other regions of the egg cortex results in a high frequency of *mitotic abnormalities* in the cleaving egg. Pluricentric mitoses are very frequent in the slightly injured eggs, leading to *aneuploidy,* a condition which, as we already know, is lethal. If an egg whose cortex has been slightly injured stops developing at the blastula stage, it becomes almost impossible to decide whether the arrest is due to purely cytoplasmic or to nuclear (aneuploidy) damage. Experiments in which either the dorsal cortex (grey crescent) or the ventral cortex had been injured prior to microinjection of labelled uridine as a precursor of RNA synthesis have given a clear answer. RNA *synthesis always runs parallel with development.* If the development stops early, at the blastula stage, RNA synthesis will be almost negligible; if abnormal embryos develop, there is a close relation between deficiencies in morphogenesis and decreased RNA synthesis. Such cases of early arrest of development or abnormal embryogenesis are much more frequent after injury of the dorsal than the ventral cortex of the fertilized egg.

The analysis becomes much easier at the *gastrula* stage, since the dorsal side is now very well marked by the presence of the dorsal lip of the blastopore. Cytochemical studies, which have been confirmed by biochemical analyses, have shown that the RNA content of the dorsal lip is higher than that of the ventral lip. Further work, with labelled uridine as a tracer, has confirmed that nuclear RNA synthesis is stronger in the dorsal than in the ventral half of the gastrula. As one would expect, protein synthesis is also higher in the dorsal than in the ventral half of the gastrula. The first contains less yolk, but more ribosomes and polysomes than the second.

Similar, but less extensive, studies indicate that the same situation prevails in the embryos of the other vertebrates (except perhaps the mammals). The *Antero-posterior* (cephalocaudal) gradient in RNA distribution corresponds to the initial polarity gradient of the fertilized egg; later on, a *dorsoventral* gradient of RNA synthesis appears and superimposes itself upon the initial antero-posterior gradient. This second gradient becomes clearly visible at gastrulation only. Morphogenesis apparently results from the interaction between these two gradients in RNA distribution and synthesis. They are indeed identical with the morphogenetic gradients of experimental embryology.

It is, of course, very important for molecular embryologists to know more about the *nature of the RNAs* which are synthesized during gastrulation and neurulation. An important part of these newly synthesized RNAs is certainly tRNA and rRNA. The nucleoli appear shortly before gastrulation and, as one would expect, their reappearance is linked to the synthesis of nucleolar rRNA precursors and cytoplasmic rRNAs. Gastrulation is thus the stage in which new ribosomes are produced. There is good evidence for the view that the syntheses of the 28 S and 18 S rRNAs, as well as that of the ribosomal proteins, are coordinated. When rRNA synthesis is derepressed in the gastrula, because the nucleolar organizers begin to work again, the synthesis of the ribosomal proteins begins in the cytoplasm.

Xenopus embryos have played a very important role in proving that the nucleolar organizers are directly involved in the synthesis of the high molecular weight (28 S and 18 S) rRNAs. An interesting mutation in this species consists in the selective loss of the nucleolar organizers as the result of a chromosomal deletion. Normal embryos and adults possess two nucleolar organizers and, in general, have two nucleoli. If one nucleolar organizer has been lost by deletion, only one nucleolus is present; development is nevertheless normal. If one makes crosses between such uninucleolate adults, there will be no nucleolar organizer at all, as expected from Mendel's laws, in 25% of the offspring. These *anucleolate* (*nu-o*) mutants are lethal and die at the larval feeding stage; they do not synthesize measurable quantities of 28 S and 18 S rRNAs; and they are also deficient in the synthesis of ribosomal proteins. This definitely proves that the nucleolar organizers are directly responsible for the synthesis of all the ribosomes which are made during embryonic development. The fact that development can proceed until a rather late larval stage further shows that the oocyte has already built up a store of ribosomes, which is sufficient to reach, without any new ribosomal RNA synthesis, a stage of development in which the tadpole is able to feed.

We do not know why rRNA synthesis is repressed during cleavage and initiated at the early gastrula stage. It is likely that cytoplasmic factors are involved in the control of rRNA synthesis. It has been reported that cleaving amphibian eggs contain a soluble factor which selectively inhibits rDNA transcription in the nucleolar organizers. However, the experimental evidence for the existence of such a specific repressor of rRNA synthesis is not as strong as one might wish.

The "midblastula transition" is, as we have seen, important for the initiation of *small RNA* (tRNAs, 5S rRNA, 7S RNA, snRNAs) synthesis: this is the stage in which, after 12 fast cleavages, a normal cell cycle (with G_1 and G_2 phases) becomes detectable and in which a critical ratio between nuclear and cytoplasmic volumes is established. The experimental evidence presently available suggests that nuclear DNA reacts, during cleavage, with a cytoplasmic suppressor of RNA synthesis. The store of this suppressor would be exhausted at the midblastula stage and transcription would begin when nuclear DNA is no longer saturated with this still hypothetical suppressor.

We already know that amphibian eggs possess a very large store of maternal mRNAs (between 10,000 and 20,000 distinct species). As in sea urchin eggs, these mRNAs are unusual in that they contain transcripts of repetitive sequences: they are more similar to heterogeneous nuclear RNAs than to typical cytoplasmic mRNAs. Synthesis of mRNAs, in particular, those which encode the histones, already begins during early cleavage in *Xenopus*. At the gastrula stage, the mRNAs which code for the ribosomal proteins become the majority of the neosynthesized messengers. This is what one should expect on pure logical grounds. During early cleavage, DNA synthesis is very fast and large amounts of histones are obviously required for nucleosome formation; the gastrula stage, as we have just seen, is characterized by the production of new ribosomes which requires the synthesis of both ribosomal RNAs and proteins. Thanks to the pioneer work of H. Denis, we have a good deal of information about *mRNA synthesis* during gastrulation. His molecular hybridization experiments have shown that new populations of mRNA molecules appear during amphibian gastrulation. Some of the new mRNAs are stage-specific: they can no longer be detected at later stages of development and it is possible that they are involved in the complex cell movements which are characteristic of gastrulation. For later stages of development a general rule seems to emerge: new, but unstable mRNA molecules are synthesized before each organ differentiates. When organogenesis is over and the cells begin to undergo cytological differentiation (for instance, when neurones form in the still undifferentiated neural tube), more stable mRNA molecules are produced.

Recent studies have shown that specific mRNAs are indeed synthesized before visible organogenesis: skeletal muscle and cardiac actin mRNAs, first appearing during gastrulation in *Xenopus,* are strictly localized in the equatorial regions which will give rise to the muscles (somites). These experiments show that the genes coding for the contractile proteins specific of muscle tissue are turned on at a precise stage of development and in a limited region of the embryo. Selective activation of the actin genes precedes by several hours the appearance of the somites and can therefore be taken as a molecular marker of muscle determination.

During *neural induction,* synthesis of *DNA* (as evidenced by increased mitotic activity), as well as of RNA and proteins, markedly increases. New species of mRNAs are produced at that time by the nuclei and their synthesis can be induced, in isolated fragments of ectoblast, by addition to the medium of Tiedemann's neuralizing and vegetalizing protein factors. Neural induction thus seems to be accompanied by a selective derepression of a set of genes, leading to a burst of new mRNA and protein synthesis. Unfortunately, the problem of neural induction has not yet attracted the interest of molecular biologists. However, with the now available methodology, it should be possible to precisely characterize the messengers and the proteins which are synthesized during induction. This would give us important clues on the mechanisms of this all-important, but still poorly understood, process.

4 The Links Between RNA Synthesis and Morphogenesis

Many experiments have confirmed, in an indirect, but convincing way, the obtained conclusion that there is a *very close correlation between RNA synthesis and embryonic development.* Only the most important ones will be briefly described here.

For instance, if *ribonuclease,* the enzyme which breaks down RNA, is injected into amphibian eggs, development stops already during cleavage; if RNA is added to ribonuclease-treated eggs, a nervous system can form.

Actinomycin, which does not inhibit the cleavage of amphibian eggs, blocks their gastrulation; at later stages, it produces microcephalic or even almost undifferentiated embryos. Certain organs, however, are more resistant to the drug than others. For instance, ears and kidneys can form in embryos which are almost devoid of a nervous system. This finding is in agreement with the idea that some of the mRNAs, which are involved in embryonic differentiation, are more stable than others. Actinomycin would inhibit the synthesis of the unstable mRNAs required for nervous system organogenesis and would not act upon the more stable, pre-existing mRNAs present in the ear and kidney areas. Other experiments with actinomycin have confirmed that induction really is a derepression. If an organizer is treated with actinomycin and placed in contact with a normal ectoblast, neural induction occurs. If, on the other hand, an actinomycin-treated piece of ectoblast is joined with a normal organizer, there is no neural differentiation at all. The ectoblast is thus unable to respond to the inducing stimulus if it cannot synthesize RNA. The organizer, on the contrary, can retain its inducing capacities in the absence of RNA synthesis. This finding is in good agreement with the already expressed view that the inducing activity is probably due to the presence, in the dorsal lip, of pre-existing proteins. Experiments with other inhibitors of RNA synthesis, e.g. *α-amanitin,* have led to similar conclusions.

Experiments with the protein synthesis inhibitor, *puromycin,* have been less extensive; they have shown that microcephaly is frequent in the treated embryos, as expected. The development of the eyes and brain requires a large increase in the number of the cells, which obviously requires DNA and protein synthesis. Of some interest is the fact that the *competence* of the ectoblast cells toward the vegetalizing protein factor of Tiedemann is retained for a longer time if the cells have been pretreated with puromycin. The factors that are responsible for the normal loss of competence of the ectoblast, during ageing of the gastrula, are still totally unknown. The experiments with puromycin suggest that the appearance of new proteins, leading to a more differentiated state, might be responsible for this loss of competence.

Finally, experiments on centrifuged and treated eggs, which cannot be described here, fully agree with the conclusion that morphogenesis, in vertebrate eggs, is controlled by the orderly synthesis of mRNA and proteins along the cephalocaudal and dorsoventral gradients.

5 Size and Mode of Action of the Inducing Agent

During neural induction, there is a very close contact between the inducing cells of the chordoblast and the reacting cells of the neuroblast. Also, in explantation experiments, if pieces of chordoblast and neuroblast are placed together, the two fragments quickly stick to each other. We shall return later to the properties of the cell membranes in the gastrula (Chap. XI). Right now, we would like to ask two questions: (1) Is there a movement of chemical substances from cell to cell during induction? (2) Is close contact between the inductor and the reacting tissue necessary for successful induction?

Experiments with explants that had previously been stained with vital dyes or treated with labelled precursors of RNA and protein synthesis clearly show that even large molecules can pass from the inducing to the reacting cells. But migration does not occur in this direction only. The dye, or the radioactive label, can move from the reacting to the inducing tissue just as easily.

Thus, the experiments do not provide any evidence for a *unidirectional* migration of macromolecules, which would be of critical importance for the success of neural induction.

We now come to the second question. In a first series of experiments, pieces of cellophane membrane were inserted, i.e. "sandwiched", between pieces of chordoblast and neuroblast. The result was the suppression of induction, and the conclusion drawn was that direct contact between the inducing and reacting cells is necessary for neural induction. The experiment allows another conclusion; cellophane membranes do not allow the passage of large molecules, such as proteins or nucleic acids, but they are very permeable to smaller molecules. Therefore, neural induction *cannot* be simply the result of an overproduction by the inducing cells of small molecules, such as CO_2, ammonia or lactic acid.

But further experiments have shown that although the second conclusion remains valid, the first one (the necessity of close contact between the two tissues) is not correct. If one uses "millipore" filters, which allow the passage of macromolecules, but not that of whole cells, instead of cellophane membranes, *transfilter* inductions can be obtained. Electron microscopy shows that the cells of both the inducing and reacting tissues send out long filaments (called *pseudopodia* or *filopodia*) inside the pores of the filter; but these filaments do not touch each other. Transfilter induction is thus due to the release of soluble substances by the inducing cells. However, a possible intervention of the glycoproteins, which form the intercellular matrix, has not yet been completely ruled out. The fact that induction can be obtained when explants of ectoblast are treated with the neuralizing or mesodermalizing proteins in solution further demonstrates that cell-to-cell contact is *not* required for successful induction. In fact, it has been shown, by cytochemical methods, that these inducing proteins are taken up by the cells by pinocytosis; some of the inducing protein is found later on in the cell nuclei, where it might derepress the chromatin.

It should be added that transfilter induction is not limited to the induction of the nervous system in amphibians. Similar observations have been made for nervous system induction in birds, for the induction of the eye lens by the optic cup, for the induction of kidney tubules, of pancreatic gland cells, etc. It seems to be a general rule that direct contact between inductor and reactor is not required for the success of induction.

All this strongly indicates an important role of the *cell membrane* (plasma membrane) in inductive processes. This conclusion is reinforced by the recent finding that induction of the nervous system by the chordomesoblast (its natural inducer) is suppressed by treatment with lectins, proteins of vegetal origin which bind specifically to the carbohydrate residues of the cell surface glycoproteins. We shall soon see that these glycoproteins also play an important role in the cell movements which characterize gastrulation.

6 Molecular Basis of Cell Movements

Cell movements play an essential part in early morphogenesis. They are required, in a particularly striking way, for gastrulation and neurulation. During the latter, the nervous system first forms a flat plate, which changes into a groove and finally into a tube.

This transformation of a neural plate into a neural tube is completely blocked by *mercaptoethanol:* this -SH containing substance, as we have seen earlier, also suppresses the formation of the mitotic apparatus during cleavage. Other substances, which affect the -SH-SS equilibrium, also inhibit the closure of the neural tube. On the other hand, if a young neurula, which has a flat neural plate, is treated with ATP, the formation and closure of the neural tube are markedly accelerated. Substances like mercaptoethanol and ATP have, of course, many biochemical side effects on the synthesis of the macromolecules. But it is probable that their main target is, as in the case of the mitotic apparatus and the cleavage furrows, the *microtubules* and the contractile *microfilaments* which have been mentioned earlier (Chap. VI). Recent work has produced strong evidence for this view. Both *colchicine* (which reacts with tubulin) and *cytochalasin B* (which affects the actin microfilaments) inhibit gastrulation and neurulation in amphibian eggs. These important morphogenetic events would thus be due to the intervention of *contractile proteins,* which could modify the shape of the cells by inducing localized contractions. The cells in the blastula have a cuboidal shape; a constriction, resulting from the contraction of a ring of microfilaments localized at the apical end of the cells, would change their shape: they would become trapezoidal. If such a change occurs simultaneously in a large number of neighboring cells, a whole population of cells would undergo the intensive cell movements, which are so characteristic of gastrulation and neurulation. Analysis, at the molecular level, of the changes in shape undergone by the cells during these early steps of morphogenesis has shown that the microtubules are responsible for the increase in length of the cells, and the microfilaments for the apical constriction which changes their shape during the formation of the nervous system in the neurula.

However, it is likely that these changes in the organization of the cytoskeleton (microtubules and microfilaments) result from signals first received at the *cell membrane* level. As shown by our colleague R. Tencer, some lectins inhibit gastrulation, while others – which bind to other carbohydrate residues – are inactive. Amphibian gastrulation is also inhibited by tunicamycin, a drug which inhibits protein glycosylation. The lectins also affect the shape and motility of dissociated gastrula cells. An important control mechanism might be the availability of free calcium ions, which play a crucial role in the polymerization and depolymerization of the microtubules and microfilaments. Unfortunately, we still do not know how the signals received at the cell membrane level are transmitted to the cell interior, where they may affect the initiation of DNA synthesis as well as the organization of the whole cytoskeleton.

While changes in cell shape are apparently sufficient to explain the impressive deformation of a neural plate into a neural tube, morphogenetic movements are also involved: a very important factor is an *oriented migration* of the cells that can be controlled by different mechanisms (e.g. chemotaxis for germinal cells). In the particular case of gastrulation, oriented migration is controlled by *contact guidance:* large groups of cells follow oriented fibrils present in the *extracellular matrices* (ECM).

The extracellular matrix is an intricate meshwork of large macromolecules: *hyaluronic acid* (a large glycosaminoglycan), *proteoglycans* (macromolecular associations of covalently linked proteins and glycosaminoglycans) and protein fibers, mainly *collagen* and *elastin* fibers. Cell adhesion to this matrix is mediated by specialized glycoproteins. Fibronectin (FN), for instance, a major ECM component, binds specifically to a cell membrane receptor at one end and to collagen at the other. It promotes both adhesion and migration of cells in many different systems and is associated with basement membranes, thus providing pathways for migratory cells.

It was shown recently that in amphibian eggs, FN is synthesized during cleavage on a stored maternal mRNA and that it accumulates specifically on the blastocoele roof where the mesodermal cell will migrate during gastrulation. The essential role played by FN in morphogenesis is demonstrated by the fact that injection of antibodies against FN into the blastocoele cavity inhibits gastrulation.

Morphogenetic cell movements thus result from complex, and still poorly understood, interactions between ECM components, plasma membrane and cytoskeleton. Further research in this field should lead to interesting and important results.

CHAPTER IX

Molecular Embryology of Mammals

1 The Biology of Mammalian Sperm

1.1 Male Sex Determination in Mammals

Spermatogenesis in mammals follows the same course as in all metazoans (see Chap. IV). If one excepts a few species, the male is genetically characterized by the presence of two different sex chromosomes, called X and Y, and is the *heterogametic* sex which produces, as a result of meiosis, two different kinds of gametes, one carrying the X and the other the Y heterochromosome. In contrast, all the oocytes are of the XX type, the females being the *homogametic* sex in mammals. After fertilization, half of the zygotes will inherit a Y heterochromosome, which will be responsible for the development of male gonads (testes); in its absence, the so-called indifferent gonads progressively become ovaries.

How the small Y chromosome is able to direct the complex development of testes is a fascinating problem since it offers a unique opportunity to correlate a complex morphogenetic process with a well localized determinant, Tdy (for testicular determinant located on the Y chromosome). However, genetic analysis has shown that other genes, located on different chromosomes, are also involved in the early steps of testicular differentiation; Tdy might be only a regulatory gene. Recombinant DNA technology will probably enable us to solve this problem in the near future.

The molecular mechanisms controlling testis differentiation are not yet understood. The only gene product that unequivocally depends on the expression of Tdy is the H-Y antigen, a protein found on the surface of all male cells, and which can be detected as early as the 8-cell stage in the mouse. Incidentally, this provides a hope for sex selection at very early stages of development since antisera against H-Y recognize only male embryos. The direct involvement of the H-Y antigen, in testis primary differentiation, has been strongly suggested: in vitro, testicular organization is inhibited by antisera against H-Y. Furthermore, addition of free H-Y antigen induces the transformation of fetal presumptive ovary into testes.

This testicular differentiation triggers a cascade of events: two kinds of embryonic hormones are synthesized. One of them is the "Müllerian Inhibitory Factor" (MIF), which is a glycoprotein responsible for the regression of the Müllerian ducts which are conserved in females (as oviducts, uterus and part of the vagina) and disappear in males. The others are the androgens which stabilize the Wolffian duct and its derivatives (epididymis, vas deferens and seminal vesicles) and trigger the formation of the external genitalia (penis and scrotum) (Fig. 55).

Differentiation of the somatic cells in the male gonad is independent of the presence of germ cells. In sharp contrast, the somatic part of the fetal gonad plays an essential role in the processes of meiosis and gamete production: the germ cells already begin meiosis within the fetal ovaries, while the male germ cells do not enter meiosis before puberty, when the production of male hormones has reached a critical threshold.

This duality could be explained by the secretion of specific substances by the gonads. A meiotic-inducing substance (MIS) would be involved in the initiation of meiosis in the female, while an antagonistic, male-specific, meiotic-preventing substance (MPS) would be keeping the germ cells inactive until puberty.

1.2 Mammalian Spermatozoa from Testis to Fertilization

Spermatogenesis leads to a huge production of spermatozoa from a stock of spermatogonia. It takes place in the *seminiferous tubules* of the testis (Fig. 55) where mitotic proliferation requires a finite number of divisions characteristic of each species (six in the rat, for instance); it is followed by meiotic divisions and spermiogenesis (Chap. IV). However, in contrast to spermatozoa of most invertebrates and low vertebrates, the testicular spermatozoa of mammals are totally unable to fertilize eggs. Two essential modifications of the spermatozoon are re-

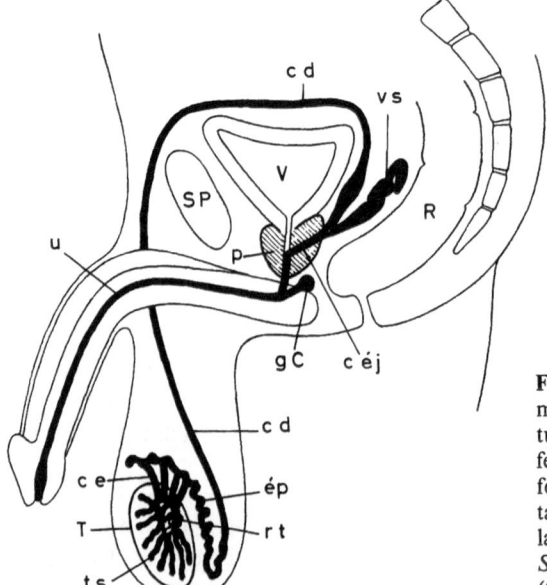

Fig. 55. The male sexual organs in man. *T* testis; *ts* seminiferous tubules; *rt* rete testis; *ce* vasa efferentia; *ép.* epididymis; *cd* vas deferens; *vs* seminal vesicle; *p* prostata; *gC* Cowper's gland; *céj* ejaculatory canal; *u* urethra; *V* bladder; *SP* symphysis pubica; *R* rectum. (Courtesy of Prof. J. Mulnard)

quired to ensure fertility. The first takes place during its journey in the epididymis (Fig. 55), the second in the female genital tract. They are called respectively, *epididymal maturation* and *capacitation.*

Epididymal Maturation: Factors responsible for the functional modification of the spermatozoa are secreted by the epithelium of the epididymis which responds to testosterone stimulation by an overall increase in DNA, RNA and protein synthesis. The journey of the spermatozoon through the epididymis, which lasts about 2 weeks, causes a complete reorganization of the *sperm plasma membrane* as shown by changes in surface charge, reduction in the surface sulfhydryl groups, changes in lipid composition and in membrane ATPase activity, appearance of new antigens, etc.

In addition, there is a shift from oxidative metabolism to glycolysis and fructose is used instead of glucose as the main substrate. The main consequence of these changes for the spermatozoon is the acquisition of flagellar motility.

As a result of this maturation, sperm taken from the *cauda epididymis* (the distal part of the epididymis) is now able to fertilize eggs after insemination. A second wave of membrane modifications (coating with several seminal plasma proteins and glycoproteins), which takes place at ejaculation, is thus not essential for fertilization per se; it seems to act as a protective mechanism for the ejaculated spermatozoon.

Capacitation: The discovery in 1951 by Austin and Chang, that neither epididymal nor ejaculated sperm is able to fertilize eggs unless it stays for several hours in the female genital tract, was the starting point of numerous papers dealing with the process called *"capacitation"* by Austin. This physiological change prepares the sperm for the *acrosome reaction* (see Chap. IV) and, subsequently, for fusion with the egg membrane. The earliest event during capacitation is a change in the motility pattern of the spermatozoa: they progressively acquire a characteristic vigorous whiplash-like beating of the flagellum. The term "hyperactivation" has been proposed to describe this movement.

The possibility of obtaining capacitation in vitro in several species has cast some doubts on the existence of specific "capacitation-factors". However, in hamster sperm, two main activities allow one to distinguish between two kinds of factors. The first factor is responsible for the hyperactivation of flagellar beating. This *"motility factor"* is a β-amino acid, taurine or hypotaurine, present at high concentrations in both male and female genital tracts. The second factor, which is responsible for the acrosome reaction, is simply serum albumin; its action is enhanced by catecholamines.

Capacitation is a multistep phenomenon, which includes removal or modification of the sperm surface coat. Among the removable constituents of the coat, are lipoprotein vesicles. When these vesicles are isolated from seminal fluid, where they are abundant, and added at high concentration to the capacitation medium, they reverse the whole process and are therefore referred to as *"decapacitation factors"* (DF).

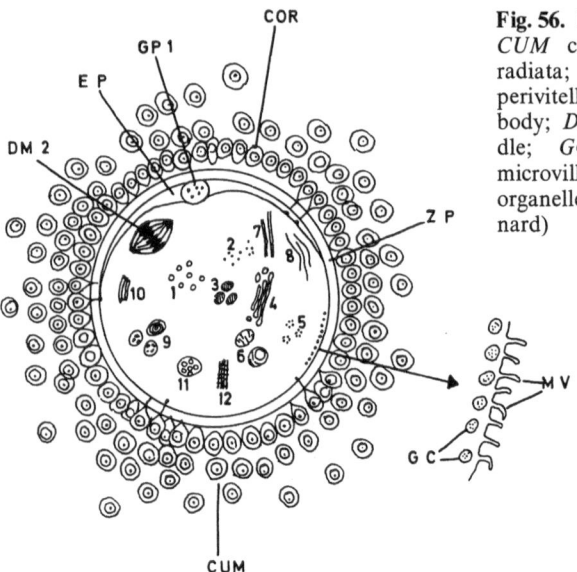

Fig. 56. Mammalian ovulated egg. *CUM* cumulus cells; *COR* corona radiata; *ZP* zona pellucida; *EP* perivitelline space; *GP1* first polar body; *DM2* second maturation spindle; *GC* cortical granules; *MV* microvilli. *1–12* Various cytoplasmic organelles. (Courtesy of Prof. J. Mulnard)

How all these agents interact to finally cause the fusion between plasma and outer acrosomal membranes (acrosome reaction) is still hypothetical, despite the fact that much more attention has been devoted to the acrosome reaction in mammals than in any other zoological group. Is is, however, certain that, as in sea urchins (see Chap. V), Ca^{2+} fluxes are involved. It has been proposed that calcium ions actively enter the space between the plasma and external acrosomal membranes where they activate membrane phospholipases with the release of free fatty acids and lysophospholipids. Such a membrane destabilization creates conditions favorable for fusion of the two membranes.

The acrosome reaction takes place in the vicinity of the *zona pellucida,* the glycoprotein coat surrounding the egg. It allows both soluble and membrane-bound acrosomal enzymes to reach their substrates. The best known of these enzymes, hyaluronidase, enables the spermatozoon to cross a first vestment, the surrounding cumulus cells and the *corona radiata* (Fig. 56). A second sperm-specific enzyme is acrosin, a trypsin-like proteinase; it was first thought to be responsible for the *zona pellucida* penetration, but it has a more restricted, although essential, role: the binding of the spermatozoon to the *zona*. The penetration of the zona is apparently ensured by other enzymes such as arylsulfatase, β-N-acetylhexosaminidase and hyaluronidase. In addition, a mechanical cutting through the *zona pellucida* by flagellar beating is not unlikely.

The specific recognition between a spermatozoon and an egg of the same species is provided by the presence, in the zona pellucida, of a glycoprotein which displays the properties of a *"zona receptor"* for a sperm glycoprotein which corresponds to the "bindin" of sea urchin spermatozoa (see Chap. V). Binding of a reacted sperm to the plasma membrane of a ripe oocyte is much less specific.

Advantage has been taken of this fact for fundamental studies on fertilization and for developing an assay for human fertility, in which zona-free hamster oocytes are fused with human spermatozoa.

2 Molecular Embryology of Mammalian Eggs

Eggs of mammals, except those of Monotremata, are among the smallest in the animal kingdom (60 to 150 μm in diameter). This is related to their very specialized pattern of development which involves the differentiation of a unique organ, the *placenta:* it provides the maternal source of nutrients needed to ensure the major part of embryonic development. The only period of independent existence of the eggs is that which elapses, during a few days, from fertilization to implantation. During this "preimplantation period", the fertilized egg undergoes cleavage and blastulation which results in the establishment of two distinct cell populations within the blastocyst (see Figs. 1 and 62). The external cell layer is the *trophectoderm* which will differentiate into invasive trophoblast and part of the placenta; inside is the *inner cell mass* (ICM) which will develop into the fetus and extraembryonic membranes (amnion, yolk sac, allantois).

One should keep in mind that there is a great difference between the cleavage pattern in mammals and in invertebrates and lower vertebrates: in the latter, all the blastomeres contribute to the development of the embryo, whereas only a few cells of the inner cell mass (only 3 of 30 in the mouse, for instance) will give rise to the whole fetus.

The molecular analysis of the very early stages of mammalian development has started recently due to the development of in vitro culture procedures and of microanalytic and micromanipulative techniques. Most of our knowledge on the biochemistry of oogenesis, oocyte maturation, fertilization and blastocyst formation comes from experiments performed on a limited number of species, mainly mouse and rabbit. Many problems concerned with human fertility definitely require research with human material. Two main sources are now available owing to recent clinical advances: "spare" embryos from in vitro fertilization (IVF) and embryo transfer programs, and oocytes given by volunteers undergoing sterilization. Such studies have already been initiated.

2.1 Oogenesis

Mammalian oocytes do not accumulate a large quantity of reserve materials and are therefore called *alecithic.* However, the eggs of the carnivora contain so-called "yolk-like particles" in their cytoplasm which have been characterized as glycolipid inclusions. They are therefore chemically different from the yolk platelets of lower vertebrates.

However, during oogenesis, the oocytes undergo a significant increase in size during the first half of follicle growth. The nuclear genes are actively transcribed during this cytoplasmic growth period, resulting in an accumulation of both ribosomal and messenger RNAs correlated with the respective levels of RNA polymerase I and II activities. In full-grown oocytes, uridine incorporation into RNA becomes very low and RNA polymerase activity is no longer detectable.

In rodents, most of the newly synthesized rRNA is accumulated into proteinaceous cytoplasmic superstructures called *cytoplasmic lattices.* These ribosomal storage forms, which do not participate in protein synthesis, are progressively disintegrated after fertilization. However, in other mammalian species, no storage form of ribosomes has been observed. The messenger RNAs, which represent about 10% of the total accumulated RNAs, are very stable and are stored in the cytoplasm; they have no specific localization.

All these results clearly indicate that as in other organisms, the mammalian oocyte is preparing the independent existence of the early embryo by the accumulation of a store of ribosomes and maternal mRNAs.

The pattern of protein synthesis changes according to the growth phase of the oocytes. For instance, the growing mouse oocyte builds up its own *zona pellucida* from three newly synthesized major sulfated glycoproteins; one of which is the already mentioned sperm receptor. Some of the developmental changes in protein synthesis are closely correlated with the acquisition of the competence to resume meiosis at maturation.

The pattern of energy metabolism is also greatly modified during the growth phase of the oocytes. They progressively lose the ability to produce CO_2 from glucose and lactate, whereas the production of CO_2 from pyruvate increases logarithmically. Pyruvate is thus the main energy substrate for large oocytes, zygotes and even early cleaving eggs.

2.2 Maturation (Resumption of Meiosis)

Oocyte maturation normally takes place at the time of ovulation and is therefore under pituitary control. The model of molecular events which underly maturation in the *Xenopus* oocyte (see Chap. IV) seems to apply equally well to mammalian oocytes: fast Ca^{2+} burst, drop in cAMP content, dephosphorylation of a "maturation protein", accumulation of MPF and, finally, germinal vesicle breakdown and chromosome condensation, resulting from an increase in protein kinase activity have all been reported.

But, in contrast to amphibians and sea urchins, mammalian oocyte maturation is not directly induced by hormonal stimulation. It is known, since the pioneer work of Pincus and Enzmann in 1935, that "spontaneous maturation" can be induced in vitro by simple mechanical release of the oocytes from their follicle. This finding has led recently to the conclusion that, in the ovary, the follicular oocyte is prevented from undergoing maturation by a follicular fluid inhibitor. Its precise nature is still controversial. It maintains a high level of cAMP which is responsible for the inactivation of a "maturation protein". Meiotic resumption thus depends on the release of this maternal inhibition, initiated by a luteinizing hormone discharge.

As already mentioned, RNA synthesis is at a very low level in full-grown oocytes. Recent experiments have clearly shown that an early transcriptional event is required for the resumption of meiosis: α-amanitin, at concentrations known to suppress RNA polymerase II activity, inhibits germinal vesicle breakdown (GVBD), if it is added immediately after the release of the oocyte from its follicle (but not later).

An early inhibition of RNA synthesis also suppresses the qualitative changes in protein synthesis which occur during maturation. This supports the idea that newly synthesized proteins are actively involved in meiotic resumption and that they are translated from neosynthesized mRNAs at the beginning of maturation. Two different sets of newly synthesized proteins can be detected: early proteins, already synthesized before GVBD, and late proteins resulting from the mixing of nucleoplasm and cytoplasm. The identity of these proteins is still poorly known. Among the former, Wassarman and Letourneau have characterized phosphorylated histone-like proteins that migrate into the nucleus, which might be constituents of the "chromatin condensation factor" responsible for meiotic and mitotic chromosome condensation. The existence of this factor has been demonstrated by Balakier and Tarkowski, who induced chromatin condensation and resumption of meiosis in small mouse oocytes as well as premature chromosome condensation (PCC) in mouse blastomeres, by fusing young oocytes (or blastomeres) with maturing oocytes.

2.3 Preimplantation Period

As in other zoological groups, the newly fertilized mammalian egg awakes and undergoes cleavage, but at a very slow pace. Mouse eggs, for instance, take almost 3 days to reach the fifth cell cycle. The morula differentiates into a blastocyst (see Fig. 1) when most of its cells have achieved their fifth cycle, i.e. when the embryo is made of about 32 cells. The cell cycles in the preimplantation embryo are characterized by the near lack of a G_1-phase and by a 6–7-h-long S-phase.

The whole enzymatic machinery needed for deoxyribonucleotide synthesis is active at all preimplantation stages. DNA polymerase α activity, responsible for semiconservative DNA replication as in sea urchin and amphibia, remains constant from the 1-cell to the 8-cell stage. Beyond that stage, it increases severalfold. The mouse zygote has also the capacity to repair DNA damages induced by chemicals or irradiation; this involves several enzymatic activities, including that of DNA polymerase β. On the other hand, neither the total number of mitochondria, nor mitochondrial DNA (which represents about one-third of the total DNA content of the unfertilized mouse egg) increase in preimplantation embryos.

Fertilized and cleaving mouse eggs display a high DNA methylase activity which is responsible for strong methylation of the DNA cytosine residues. An interesting consequence is the repression of the viral genome if a retrovirus is introduced into mouse zygotes and morulae. DNA methylation is related to gene expression in many biological systems, as we have seen in Chap. III, and seems to be one of the essential control mechanisms of gene expression in early mammalian development. It was indeed shown that the DNA of the differentiated trophoblast in the rabbit blastocyst is poorly methylated as compared to the highly methylated DNA of the still pluripotent inner cell mass.

The newly fertilized egg inherits a large supply of maternal RNAs stored during oogenesis. rRNA accounts for about 75% of the total RNA content. As a consequence of fertilization, quantitative and qualitative changes in protein synthesis take place during the first cell cycle. They result from the control of maternal mRNA translation. Total RNA, polyadenylated mRNA, ribosomal RNA and the total number of ribosomes decrease from fertilization to the late 2-cell stage, after which a marked increase in all these RNAs takes place until the blastocyst stage. The bulk of the inherited maternal RNAs has been degraded in the 2-cell stage embryos; here, it is replaced by new, embryonic transcripts. Indeed, analyses have shown that the synthesis of all classes of RNAs [rRNA, 4 S RNA, poly(A)-mRNA] can be detected in the mouse as early as the 2-cell stage, which presumably starts already soon after fertilization. The developmental importance of this RNA neosynthesis is shown by the fact that the classical inhibitors of RNA synthesis, actinomycin D and α-amanitin, inhibit both cleavage and blastocyst formation.

As a consequence of this stimulation of the embryonic genome expression, a dramatic shift in the newly synthesized polypeptides is observed during the second cell cycle: numerous new polypeptide species are synthesized on the new embryonic transcripts, while the majority of the maternal messengers, which had been used before this stage, become ineffective. However, a few masked maternal mRNA species remain available for protein synthesis at the same time. The significance of this late post-transcriptional activation of a few maternal transcripts has still to be investigated. The transition from the maternal to the embryonic control of development thus takes place at a very early embryonic stage in comparison to echinoderms (Chap. VII) and amphibians (Chap. VIII).

More discrete qualitative changes in polypeptide synthesis take place during the following cell cycles; some of them are clearly limited to either the inner cell mass or the trophectoderm at the onset of blastocyst formation. Not only stage-specific, but also tissue-specific polypeptides are identifiable. Their synthesis is an absolute prerequisite for the morphogenetic events leading to blastocyst formation. Inhibition of their synthesis, which takes place at the beginning of the fifth cell cycle, prevents blastocyst formation. Thus, both molecular and morphological differentiation seem to be under the control of a biological clock consisting in the achievement of a given number of DNA replication cycles.

3 Experimental Embryology of the Preimplantation Stages

We have seen that at the end of cleavage, the eggs of the mammals have reached the blastocyst stage. Gardner, who was the first to adapt microsurgery to the experimental study of mouse embryology, clearly demonstrated that the two cell populations of the 3.5-day blastocysts are already committed. The mechanical and immunosurgical separation of pure trophectoderm and pure inner cell mass (ICM) enabled him to transfer the two tissues separately into foster mothers. The former implants normally, induces the decidual reaction of the uterine stroma and differentiates into giant polyploid trophoblast cells. The latter (ICM) fails to implant but, if reassociated with pure trophectoderm from another origin, undergoes complete differentiation in all embryonic and extraembryonic tissues, except trophoblast (Fig. 57). According to classical embryological terminology, trophectoderm of full-grown blastocysts is thus a determined and unipotent tissue, whereas ICM is largely pluripotent since it loses only the trophoblastic potentiality.

Because of this very early divergence in determination and because of the increasing interest for mammalian embryology, the cleaving mouse egg has become a choice material for the study of cell commitment, determination and control of morphogenesis. Three main questions may be asked: How is cell commitment generated? When are the trophectodermal cells determined? Which factors control blastocyst formation (blastocystogenesis)?

Two main theories have been proposed to answer the first question. In the late 1940s and early 1950s, Dalcq proposed the existence of a bilateral symmetrical organization in the uncleaved eggs, which would result from an excentric distribution of the cytoplasmic constituents. Seidel proposed the existence, in the rabbit egg, of a specialized subcortical area organized like the *formation center*

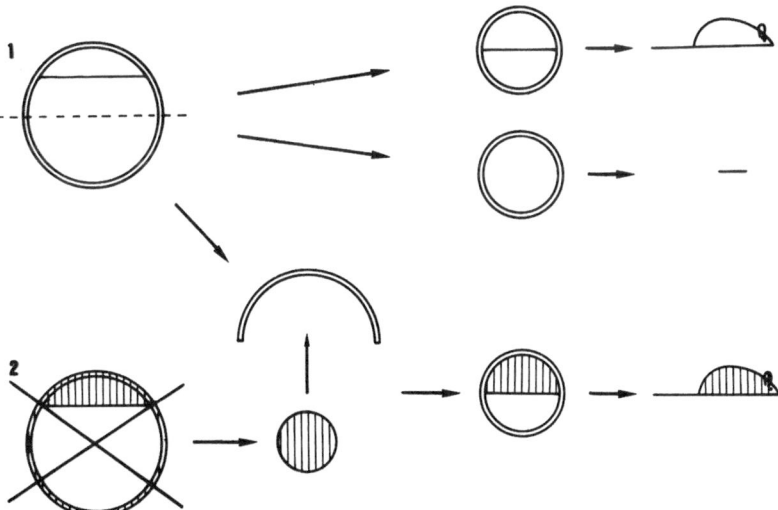

Fig. 57. Microsurgical demonstration of the respective developmental potentials of inner cell mass (*icm*) and trophectoderm in the mouse blastocyst. (According to R. L. Gardner). *1* Bisection of a blastocyst. *2* Reconstitution of mouse blastocyst from genetically different mural trophectoderm and inner cell mass. (See text for details)

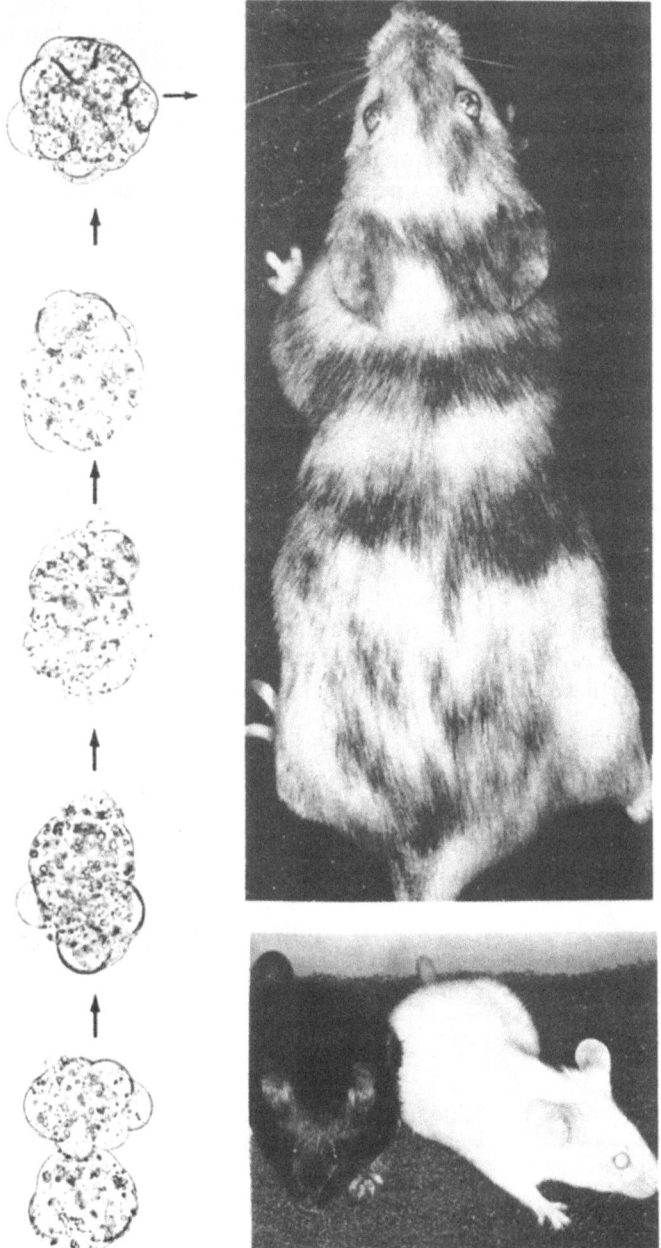

Fig. 58. Obtainment of allophenic mice: *Left, from bottom to top:* in vitro aggregation of two living cleavage stage embryos of pigmented and albino genotypes. *Upper right:* An allophenic mouse from such paired genotypes. *Lower right:* Two of the offspring of this allophenic female mouse after mating with an albino male. One comes from a genetically pigmented, the other from an albino germ cell (Mintz 1971)

(*Bildungszentrum*) of the insects, which would confer to some of the blastomeres the ICM differentiation potentiality.

This *predetermination* or *segregation* theory became very doubtful when Tarkowski (1961) and Mintz (1962) demonstrated that two genetically marked cleaving embryos made of 8 to 16 blastomeres can be aggregated together to form a single, giant tetraparental blastocyst. It can be implanted into a foster mother and it develops into a so-called *allophenic* mouse made of two phenotypically distinguishable cell populations (Fig. 58). This remarkable *regulation* capacity clearly demonstrated that mammalian eggs are of the regulative and not of the mosaic type (Chap. III). In addition, the classical experiment of Driesch on sea urchin eggs at the 2-cell stage, which have led to the concept of embryonic regulation (see Chap. III), could be successfully repeated on mouse, 2-cell-stage embryos: after separation of the two cells, each blastomere forms a small blastocyst which can develop into a normal mouse. However, when 4- and 8-cell-stage embryos are dissociated, it is observed that the later the dissociation is carried out, the more numerous are the vesicular structures made of trophectoderm only. This crucial experiment enabled Tarkowski and Wroblewska to propose in 1967, the *epigenetic* or *outside-inside theory* of primary differentiation in mammals: determination depends solely on the *position* of the cells within the embryo. The blastomeres, which are enveloped by others and thus separated from the external medium, acquire an "embryonic" determination as a result of their new microenvironment. In contrast, the peripheral cells, still in contact with the external medium, retain the basic potential of all blastomeres to become trophectoderm.

At the 8-cell stage, the mouse embryo undergoes an important morphological change, called "compaction" because adhesion of the blastomeres to their neighbors becomes so tight that the interblastomeric furrows seem to disappear (Fig. 59). This compaction is due to the formation of both tight junctions and gap junctions between the blastomeres.

It is known, at the molecular level, that compaction depends on sterol synthesis, glycosylation of membrane proteins and organization of the microfilaments (but not of the microtubules) in a network. Compaction is a Ca^{2+} dependent morphogenetic event: calcium ions play an essential role in cell membrane apposition by triggering structural modifications in a special kind of cell adhesion molecule (see Chap. XI) that has been called *uvomorulin* or *cadherin*. Johnson and co-workers have shown that at compaction, the blastomeres become polarized: they all display the *cytoplasmic polarity* already described long ago by Dalcq. In addition, there is a structural asymmetry in their plasma membrane, i.e. the portion of the embryo corresponding to the exposed surface has many microvilli and is rich in lectin receptors, in contrast to the apposed portions of the membranes. The latter displays an intense alkaline phosphatase activity which is completely absent on the exposed surfaces. As a consequence of radial mitoses, two kinds of cells are generated at the 16-cell stage: large external polar cells and small internal apolar cells (Fig. 59). The apolar cells are much more adhesive than the polar ones. Consequently, dissociated apolar cells reaggregate and recompact much more rapidly than dissociated polar cells. In contrast, aggregates

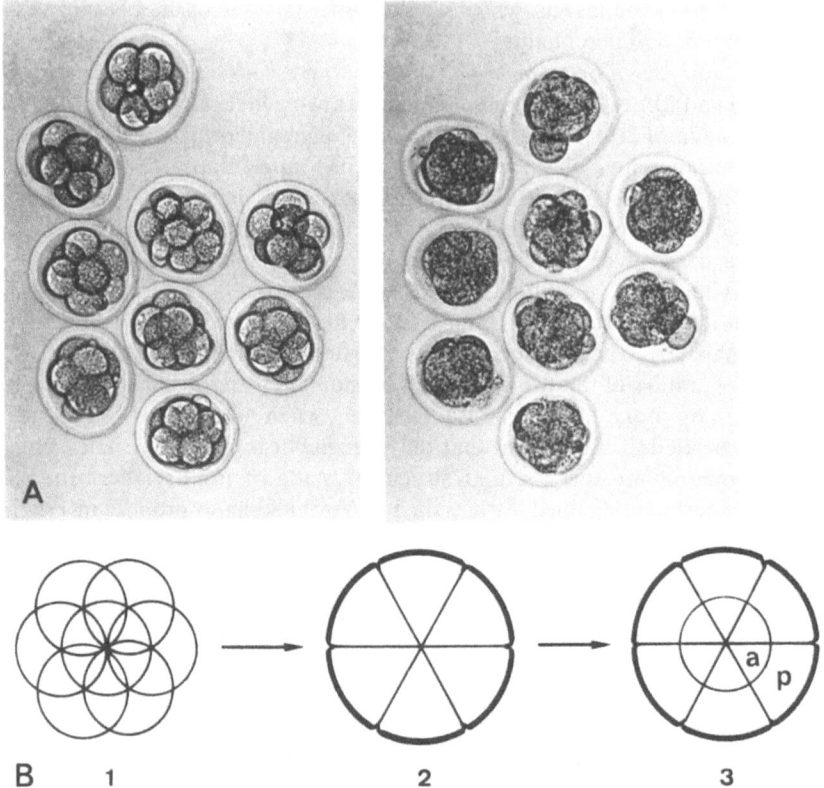

Fig. 59 A, B. Compaction and polarization in mouse eggs. **A** Eight-cell stage embryos before (*left*) and after compaction (*right*). **B** The polarization hypothesis: polarization takes place at compaction (*1* → *2*). At the 16-cell stage (*3*) apolar (*a*) and polar cells (*p*) are generated

of polar cells cavitate (i.e. form a blastocoele cavity) earlier than aggregates of apolar cells. However, both reconstituted embryos finally develop into normal blastocysts. If they are placed in contact, a large polar cell tends to engulf a small one. Thus, a polar cell will occupy an external position even if it has been placed in the center of a reconstituted aggregate made of 15 apolar cells, whereas a single apolar cell associated with 15 polar cells will be determined according to its location. These fascinating experiments suggest that as a consequence of the cellular polarization, which occurs at the 8-cell stage, a differential segregation of cytoplasmic and membrane territories takes place during the next cleavage, thus leading to early cytodifferentiation.

Regarding the second question (when are the cells determined?) it has been possible to demonstrate the totipotency of every blastomere at the 8-cell-stage. Contribution to both trophoblast and fetuses in eight reconstituted embryos was demonstrated for each blastomere of the same origin.

Fig. 60 A–C. A mouse 8-cell embryo has been injected into the blastocoele cavity of a giant blastocyst (**A**). It is surrounded by its *zona pellucida* and forms a normal blastocyst (**B**); without a *zona*, the injected embryo forms a secondary ICM only. (From Pedersen and Spindle 1980; reprinted from *Nature*, 1980, 284: 550–552 by copyright permission of Macmillan Journals Ltd.)

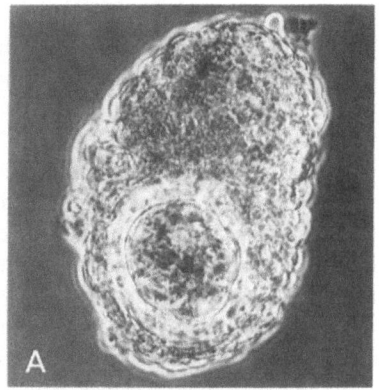

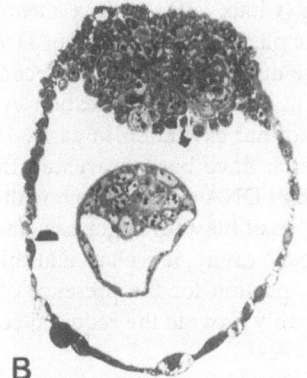

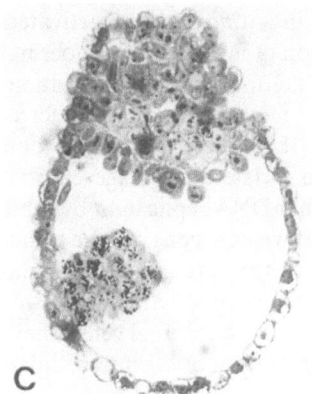

While no equivalent experiments have been performed so far at later stages, dissociation and reassociation experiments strongly suggested that at the 16-cell stage, both external and internal cells remain totipotent despite their obvious cytodifferentiation. At the 32-cell stage, the presumptive trophectodermal cells are already committed, while the internal cells remain totipotent until the mid-blastocyst stage.

It is now well demonstrated that an asymmetric cell contact is responsible for the cell polarization and, consequently, for trophectoderm determination. The ICM cells of expanding blastocysts are protected against such an asymmetric contact by trophectoderm cytoplasmic processes. One of the most convincing experiments concerning the importance of cell contacts (and not only of cell environment) in the fate of the morula blastomeres is the following: a morula was introduced inside a giant blastocyst made of eight to ten aggregated embryos. The morula was either denuded or left inside its *zona pellucida* in order to either allow or avoid cellular contacts with the host blastocyst. The two morulae displayed entirely different patterns of development, i.e. they differentiated into an inner cell mass and a complete blastocyst, respectively (Fig. 60).

A last question concerning the control mechanisms of blastocystogenesis is the following: What is the nature of the *biological clock* which triggers the initiation of *cavitation?* Blastocyst formation starts by an intracellular vacuolization which takes place suddenly in a few external cells. This is followed by the lateral fusion of these initial large vacuoles into an extracellular cavity that enlarges by exocytosis and fluid accumulation by active pumping.

This process follows the already mentioned molecular differentiation and takes place at a precise stage of cleavage (about 32 cells in the mouse). It is known that neither the cell number itself, nor the time elapsed since fertilization are the triggers for cavitation. A better candidate for such a role seems to be the number of evolved DNA replication cycles, since both molecular differentiation and intracellular vacuolization take place in embryos in which cleavage has been inhibited by cytochalasin B. This situation recalls differentiation without cleavage in a *Chaetopterus* activated egg (Chap. VII). As in *Chaetopterus,* differentiation of the mouse egg does not take place in the absence of DNA synthesis unless the fourth DNA replication cycle after fertilization has been achieved. This cycle might therefore be referred to as a "quantal replication cycle" (see Chap. XI). It is indeed possible to obtain normal cavitation in early, 16-cell-stage and even in late 8-cell-stage embryos which have been prevented from undergoing further DNA replication by inhibition of DNA polymerase α with aphidicolin.

If we now consider the ultimate step of blastocystogenesis, namely the formation of an overt extracellular blastocoelic cavity, it is clear that this step is the only one which strictly depends on compaction for the presence of tight junctions between the external cells. They probably provide the required conditions for the lateral fusion of the intracellular vacuoles.

4 Early Postimplantation Stages

Most of the data concerning early postimplantation stages result from studies made on mouse embryos, thus it would be unwise to generalize them for all mammals, although they are probably valid for most of them.

It is difficult to approach molecular and experimental embryology of postimplantation stages, because no large-scale in vitro culture methods are available so far. However, one should mention that Y.C. Hsu obtained after years of investigations, normal in vitro development of mouse embryos cultured from the 2-cell stage until the limb-bud stage, corresponding to a 11-day-old embryo.

4.1 Trophectoderm Differentiation

The only differentiation event which can be studied under in vitro standardized conditions is the formation of the primary giant trophoblast cells resulting from the outgrowth of the mural trophectoderm surrounding the blastocoelic cavity (Fig. 61); this outgrowth mimics uterine wall invasion. The differentiation of the

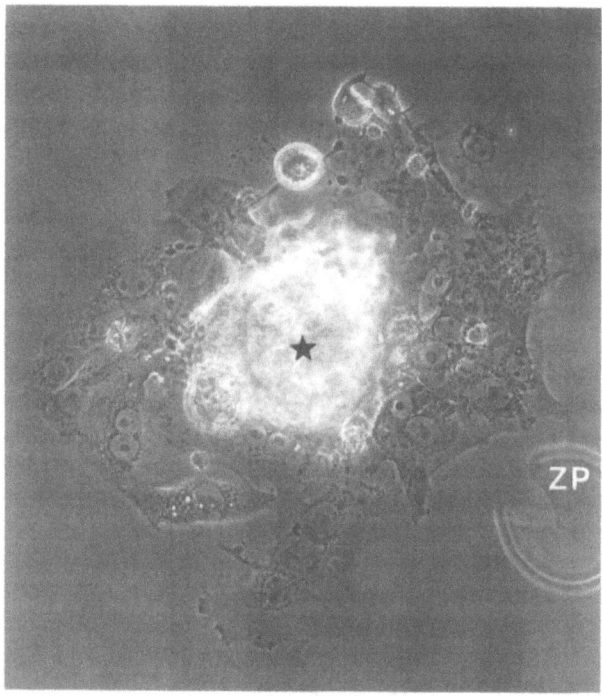

Fig. 61. Blastocyst outgrowth after culture for 70 h in the presence of fetal calf serum; it displays a well-developed ovocylinder (*star*). *ZP* empty zona pellucida

mural trophectoderm cells is characterized by DNA endoreduplication leading to polyploidization, and by the production of specialized enzymes such as Δ^5, 3β-hydroxysteroid dehydrogenase (3β-HSD) and a protease, plasminogen activator. The differentiation of the polar trophectoderm, which covers the ICM, follows a different pathway: it develops into an *ectoplacental cone* and an *extraembryonic ectoderm,* both made of mitotically dividing cells (instead of polyploid giant cells) (Fig. 62).

Blastocyst bisection and reconstitution experiments enabled Gardner to demonstrate that all trophectoderm cells have the potential to transform into giant cells; however, they are unable to undergo this kind of differentiation if they are in close contact with an inner cell mass. The latter thus displays a local inhibitory effect, which explains the growth of both the ectoplacental cone and extraembryonic ectoderm at later stages. Unfortunately, nothing is known about the biochemical mechanisms of these processes. The unique property of the trophectoderm-derived cells to undergo polyploidization in the absence of the inhibitory influence of embryonic cells has clarified the origin of the extraembryonic ectoderm, which remained controversial for a long time. When cultured in vitro or grafted on an ectopic site of a mouse (testes or kidney capsule), cells of the extraembryonic ectoderm (as well as those of the ectoplacental cone) cease im-

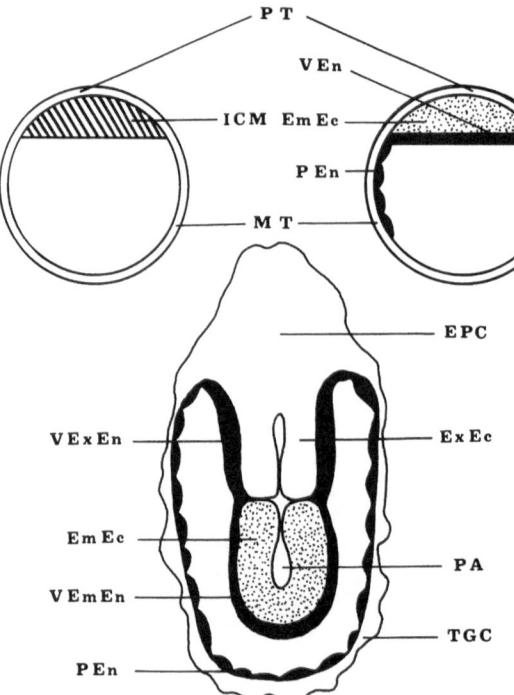

Fig. 62. Three successive stages of the development of the mouse blastocyst. *PT* polar trophectoderm; *MT* mural trophectoderm; *ICM* inner cell mass; *VEn* visceral endoderm; *EmEc* embryonic ectoderm; *PEn* parietal endoderm; *EPC* ectoplacental cone; *ExEc* extraembryonic ectoderm; *VEmEn* visceral embryonic endoderm; *VExEn* visceral extraembryonic endoderm; *PA* preamniotic cavity; *TGC* trophoblastic giant cells

mediately to divide and transform into *giant trophoblastic cells,* whereas embryonic ectodermal cells do not.

Microinjection of embryonic cells (ICM) into genotypically distinct blastocysts leads to the production of chimaeric mice. According to their determination, the injected embryonic cells contribute to the formation of various tissues in the developing chimaeric fetus. Chimaerism can be demonstrated, at the level of a single fetal tissue or membrane, by the presence of two genetically determined variants of an enzyme, glucose phosphate isomerase (GPI-Ia and GPI-Ib); the former is encoded by the homozygous Gpi-1a/Gpi-1a donor-derived cells and the second by the homozygous Gpi-1b/Gpi-1b host blastocyst-derived cells.

When either extraembryonic ectoderm or ectoplacental cone cells of 5.5-day-old embryos were injected into 3.5-day-old blastocysts, they always failed to contribute to any tissue other than the trophoblast itself, thus unequivocally demonstrating that they have undergone irreversible determination.

4.2 ICM Development

We shall only deal with the early differentiation of the inner cell mass with special attention to extraembryonic endoderm formation. Already before implantation, the ICM differentiates into a didermal embryo made of primitive ectoderm (epiblast) and primitive endoderm (hypoblast); its formation at the periphery of the ICM has been shown to be under an epigenetic, positional control (as for the trophectoderm). Cell injection experiments on 3.5-day-old blastocysts clearly demonstrated that the primitive endoderm is determined as soon as it is formed. It is strictly restricted to the endoderm layers of the yolk sac, whereas the still pluripotent epiblast cells retain the fetal and extraembryonic mesodermal potentialities, but they have lost the extraembryonic endodermal potentialities.

The following steps of extraembryonic endoderm differentiation is one of the best documented examples of genetic programming modulated by cellular interactions in mammals. The primitive endoderm spreads all over the mural trophectoderm where it becomes the so-called *parietal endoderm*. Its counterpart, still in contact with the epiblast, constitutes the *visceral endoderm*. As a consequence of the axial growth of the extraembryonic and embryonic ectoderms into an "ovocylinder", the visceral endoderm becomes subdivided into the morphologically distinguishable *visceral extraembryonic endoderm* and *visceral embryonic endoderm,* respectively (Fig. 62).

In addition to their respective morphology and location, these three primitive endoderm derivatives are clearly different from a molecular viewpoint. The parietal endoderm synthesizes plasminogen activator together with two glycoproteins (laminin and type-IV procollagen), which are deposited in a thick basement membrane apposed to the giant trophoblast cells, the so-called *Reichert membrane.* The visceral endoderm forms a thin basement membrane rich in another glycoprotein, fibronectin. The visceral endoderm also synthesizes α-fetoprotein, a protein similar to albumin, already before the extraembryonic mesoderm has been formed. However, at that stage, its expression is restricted to the visceral embryonic endoderm. Different genes are thus expressed in parietal and in visceral endoderm.

In vitro culture of isolated pieces (alone or in association) of the early ovocylinder (Fig. 62) clearly demonstrated the essential role of cell-cell interactions in the determination of these cells. It has been shown that the visceral extraembryonic endoderm has the potentiality to express the α fetoprotein gene; but it is prevented in doing so until the inhibitory influence of the underlying trophectodermal tissue (the extraembryonic ectoderm) has been withdrawn, when it is replaced by yolk sac mesoderm during normal development. Conversely, when left alone or in contact with trophectoderm-derived cells, visceral endoderm is able to differentiate into parietal endoderm; it then synthesizes large amounts of laminin and type-IV procollagen.

Several aspects of the molecular mechanisms controlling gene expression during visceral and parietal endoderm differentiation have been extensively studied on *teratocarcinoma cells* induced to differentiate in vitro. We shall come back to them in Chap. XI.

The cell lineage of the early development of the mouse is thus now well documented, however, data concerning stages beyond gastrulation are still fragmentary. This is mainly due to the lack of a suitable cell marker. Very recently, molecular embryology has come to the rescue of morphology: it has been possible to specifically identify the cells of *Mus musculus* in viable chimaeric *Mus musculus* ↔ *Mus caroli* embryos by in situ hybridization of a radioactively labelled DNA probe, specific for *Mus musculus* satellite DNA. It should now be possible to study very accurately the mouse embryonic cell lineages.

4.3 X Chromosome Inactivation

The heterochromatic body (the so-called Barr body) which is detectable in the interphase nuclei of several female mammals is, as shown in 1961 by M. Lyon, a consequence of the inactivation of one of the two X chromosomes early in embryonic development. Only one of the two X chromosomes retains its transcriptional activity and this creates, in all adult female cells, a situation equivalent to that existing in the male. In addition to the absence of production of any gene product, the "inactive X chromosome" displays two properties which make its detection possible: strongly condensed chromatin and late replication.

X chromosome inactivation is an exciting field for molecular geneticists. It is now well established that the inability of the DNA of the inactive X chromosome to be transcribed is correlated with hypermethylation of its sequences. Changes in DNA methylation are responsible for the switching off of various genes, as we have seen in Chap. III. The recent progress made in mammalian embryology has allowed the discovery of an unexpected sequence of events in X chromosome inactivation. The X chromosome of the spermatozoon being inactive, the only active X chromosome of the fertilized egg is thus the maternal X (Xm). In mouse eggs, the paternal X chromosome (Xp) is reactivated at the 8–16-cell stage; thus, both X chromosomes are active in morulae. Inactivation first takes place at the blastocyst stage, in the trophectodermal cells only; both X chromosomes are still active in the inner cell mass. A second wave of inactivation occurs in the extraembryonic endoderm at the onset of its differentiation (Fig. 62). The pluripotent epiblast still expresses the genes present in both X chromosomes, which is still the case in at least some of the ectodermal cells of 6.5-day-old embryos. The germ cells are the last to undergo X chromosome inactivation. Reactivation occurs at the onset of meiosis so that both paternal and maternal X chromosomes are active throughout oogenesis. Thus, X chromosome inactivation takes place in a recurrent way concomitantly with differentiation processes in the embryo, while reactivation occurs only in the female germ cells at meiosis.

This sequence of events has been clearly established in the mouse, due to a very sensitive assay consisting in the determination, at the level of a single embryo and even of a part of it, of the relative activities of an X chromosome-coded enzyme, hypoxanthine phosphoribosyl transferase (HPRT) and an autosomal-coded enzyme, adenine phosphoribosyl transferase (APRT). This method failed to give indications of the paternal or maternal origin of the inactivated X chromosome, but this gap has recently been filled.

The discovery of two electrophoretic variants of the X chromosome-encoded enzyme phosphoglycerate kinase (PGK-1 A and PGK-1 B) now makes the distinction between the two alleles in heterozygous female mouse embryos possible.

It was found that in the extraembryonic tissues (trophoblast and extraembryonic endoderm), a preferential paternal X inactivation occurs, whereas in all three germ layers of the fetus, in the yolk sac mesoderm and in the germ cells, X chromosome inactivation is a random process.

5 Interspecific Hybrids and Chimaeras

We cannot leave mouse embryology without mentioning a few remarkable experiments which have important implications for cell differentiation.

Even when fertilization takes place between two closely linked species by either in vivo or in vitro insemination, pregnancy is unsuccessful except for very few cases such as horse and donkey, or lion and tiger. Transfer of embryos between females of two different species is successful only in the same few exceptional cases. Failure of pregnancy after both interspecific hybridization and transfer seems to be due to a maternal immunological reaction against embryonic antigens carried by the trophoblast tissue. This idea is reinforced by the pro-

Fig. 63. Interspecific chimaera between sheep and goat. (Courtesy of Fehilly, Willadsen, Tucker 1980; reprinted from *Nature,* 1984, 307: 634–636 by copyright permission of Macmillan Journals Ltd.)

duction by *Mus musculus* females of viable *Mus caroli* offspring after transplantation of reconstituted blastocysts made of *Mus musculus* trophectoderm and *Mus caroli* ICM.

We have seen that it is possible to aggregate together two mouse morulae possessing different genetic markers (fur coat color, for instance), to culture them and to implant them into a foster mother of known genotype: one obtains "mosaic" mice, like the famous *zebra* mice shown in Fig. 58. Such mice have been called *allophenic* or *tetraparental* because each of the two fused morulae had its own father and mother. Their analysis has been important for our understanding of embryonic differentiation. Using the same experimental procedure, viable chimaeras between *Mus musculus* and *Mus caroli* have been produced by transferring the tetraparental blastocysts to *Mus musculus* recipients; the presence of trophoblast cells of maternal uterine genotype allows the chimaera to survive. More impressive are the intergeneric chimaeras created by recombination of sheep and goat blastomeres: some of the chimaeras looked like a goat with a woolly skin (Fig. 63). This was perhaps the origin of the mythological chimaeras and centaurs!

Biochemical Interactions Between the Nucleus and the Cytoplasm During Morphogenesis

We have repeatedly seen that the nucleus and the cytoplasm continuously interact during normal egg development. These interactions can be experimentally modified in two different ways. The most radical experiment is to divide the egg into two parts (merogony) and to compare the morphogenetic and biochemical potentialities of the *nucleate* and *anucleate* halves. Another approach is to modify the *nucleus*. One of the simplest ways to attain this goal is to introduce a foreign nucleus into the egg by hybridization or nuclear transplantation.

Biochemical work on anucleate fragments of eggs and on hybrids will be the main topic of this chapter. We shall, however, start with another object, the unicellular giant alga *Acetabularia,* since this organism presents a number of advantages for those who are interested in the role of the nucleus in morphogenesis and in the biochemical interactions between the nucleus and the cytoplasm.

1 Biology and Biochemistry of the Alga Acetabularia

1.1 Biological Cycle. Regeneration

The giant green alga, *Acetabularia mediterranea,* which is 3 to 5 cm long, has a single nucleus during the major part of its life cycle (Fig. 64). This nucleus, which is large and contains giant RNA-rich nucleoli, but very little chromatin, is localized at the basal part of the alga (which is called the *rhizoid*). At the apical end of its stalk, the still uninucleate alga forms an umbrella (or *cap*) which will later serve for its sexual reproduction (formation of cysts, which will produce motile gametes). The morphology of the cap is species-specific and allows taxonomists to classify the different species of *Acetabularia.* The most important of these species, for biochemists, are *A. mediterranea* and *A. crenulata,* which can easily be cultivated in large amounts in the laboratory.

Morphogenesis is affected, in *Acetabularia,* by the amount of *light* it receives. If the light supply is abundant, caps form quickly on short stalks. Reduced light supply results in the formation of very long stalks and delayed production of small caps. Absence of light completely stops growth and induces a number of degenerative changes in the nucleus (decrease in size, reduction of the dimensions and RNA content of the nucleolus); all these changes are reversible if light is again supplied to the algae. These simple experiments clearly show that the

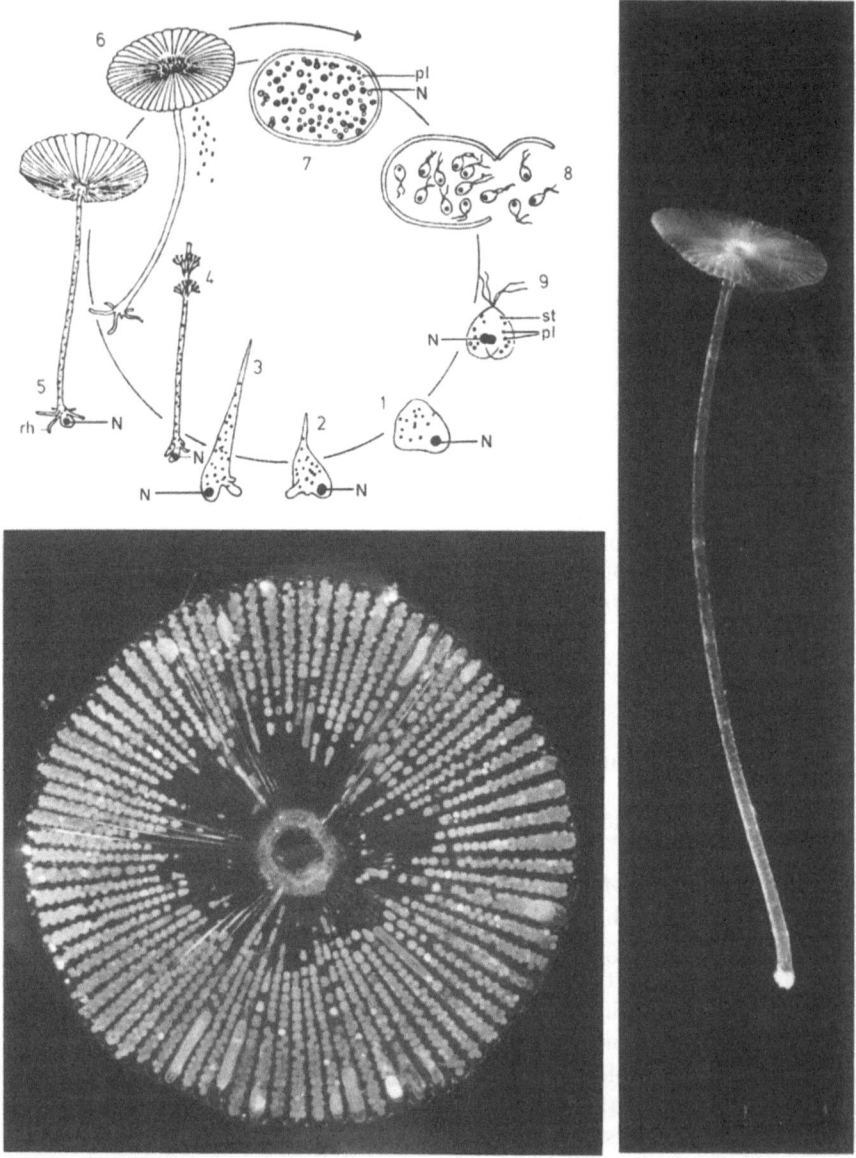

Fig. 64. *Acetabularia mediterranea. Upper left:* Life cycle of the alga. *1* zygote; *2* young growing cell; *3* slowly growing alga; *4* alga in its rapid growing phase; *5* alga with a young cap; *6* alga with a mature cap; the vegetative nucleus *N* has broken down and cysts disperse out of the cap. *7* Enlarged resting cyst; *8* germinating cyst giving rise to the gametes; *9* conjugation between two gametes. *N* nuclei; *pl* chloroplasts; *st* stigma; *rh* rhizoids (after J. Hämmerling). *Below left:* cyst formation in a cap. *Right:* whole alga

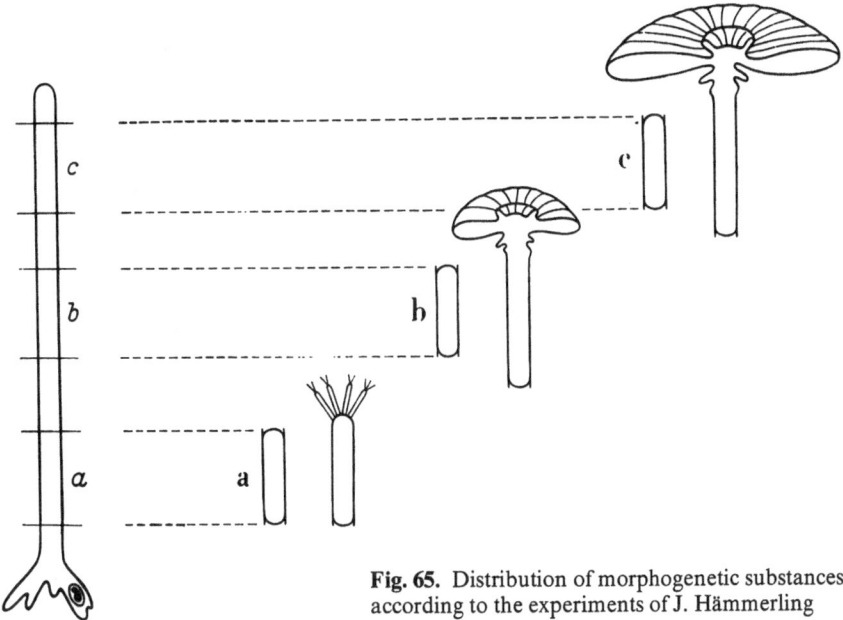

Fig. 65. Distribution of morphogenetic substances according to the experiments of J. Hämmerling

structure and chemical composition of the nucleus depend on the energy supply in the cytoplasm derived from photosynthesis.

Acetabularia has become a fascinating subject for biologists and biochemists since the fundamental work of Hämmerling around 1930. He discovered that if the alga is cut into two pieces, both the nucleate and anucleate fragments can *regenerate* and form a typical cap. Anucleate fragments can survive for several weeks and form caps, provided that the apical part of the stalk is present. The algae thus contain *morphogenetic substances,* which are distributed along an *apico-basal gradient;* as shown in Fig. 65, basal fragments, although they are close to the nucleus, hardly regenerate and never form caps. Nevertheless, the morphogenetic substances are *gene products* and originate from the *nucleus.* This has also been demonstrated by Hämmerling, who succeeded in grafting nucleate fragments of A. *mediterranea* on anucleate stalks of A. *crenulata* and vice versa. The result is, in general, the formation of a "hybrid" cap, which soon degenerates and is replaced by the typical cap of the nucleate fragment. There is a kind of "struggle" between the pre-existing morphogenetic substances present in the anucleate half of one species and the newly formed substances produced by the nucleus present in the nucleate half. The battle is ultimately won by the nucleus. The reasons why the gradient in the distribution of the morphogenetic substances decreases from the apical toward the basal end of the stalk and not, as one would expect, from their nuclear origin, in the opposite direction, are not clear. Experiments by Russian workers have shown that it is possible to displace, by centrifugation, almost the whole content of the alga toward the apical or the basal end of an anucleate fragment. Nevertheless, the polarity remains unchanged and the cap forms at the initial apical end. Thus, it is probable that polarity is linked to the properties of the cell membrane and cell wall, which are different at the apical end, where growth occurs. There is an accumulation of RNA, proteins, glyco-proteins and polysaccharides at this apical end. It is possible, but this remains a hypothesis, that the membrane contains specific receptors for the morphogenetic substances; such receptors would be accumulated at the apical end of the stalk.

1.2 Morphology of the Cytoplasm and the *Nucleus*

The algae are protected by a thick extracellular wall made of polysaccharides; it is in close contact with the plasma membrane which limits the cytoplasm. The center of the alga is occupied by a large, turgescent, acidic vacuole. The cyto-plasm is reduced to a thin sheet, compressed between the extracellular mem-brane and the central vacuole. Golgi bodies, which are accumulated at the grow-ing apical end of young algae where they participate in the formation of the poly-saccharide cell wall, and mitochondria are abundant in the cytoplasm. But the most obvious cell organelles are the *chloroplasts;* they are ovoid and often con-tain inclusions of the polysaccharides, which they have synthesized by photosyn-thesis. If one observes an *Acetabularia* under the light microscope, one can see that the chloroplasts are moving continuously on cytoplasmic strands; this move-ment is called *cyclosis.* Cytoplasmic contractile proteins (actin, myosin) are in-volved in cyclosis, since chloroplast movement quickly stops if the algae are treated with the actin-binding drug, cytochalasin.

The large (300 μm in diameter) *"vegetative" nucleus* of the alga contains, ac-cording to the *Acetabularia* species, a single large ribbon-shaped nucleolus or several smaller nucleoli (Fig. 66). The nucleoli are very rich in RNA and are the site, as we shall see, of intensive rRNA synthesis. It has been impossible for a long time to demonstrate cytochemically the presence of DNA in the huge veg-etative nucleus. On the other hand, DNA is easily detectable in the much smaller nuclei of the gametes and zygotes. Recent studies have shown that the vegetative nucleus contains *lampbrush chromosomes* similar to those of amphibian oocytes: 20-μm-long, DNA-containing loops extend in the abundant nuclear sap from condensed chromomeres. The nucleus, despite its large size, is diploid and con-tains a little more DNA than the theoretical diploid value, which is due, as we shall see, to amplification of extrachromosomal (nucleolar) genes. When the cap reaches its full size, the nucleus shrinks, the nucleoli regress and a huge in-tranuclear spindle is built up. Meiosis occurs at this time. The numerous small daughter nuclei, which move toward the cap and invade it, are haploid. In the cap, groups of such "secondary" nuclei, which have been repeatedly dividing, become surrounded by a polysaccharide membrane: the cap is now subdivided in cysts (Fig. 64); each of them contain many haploid nuclei. Finally, the cysts are released in seawater and give rise to a swarm of flagellated haploid gametes.

As one can see, there is a remarkable similarity between *Acetabularia* and eggs from animals so far as the behavior of the nucleus is concerned: the vegeta-

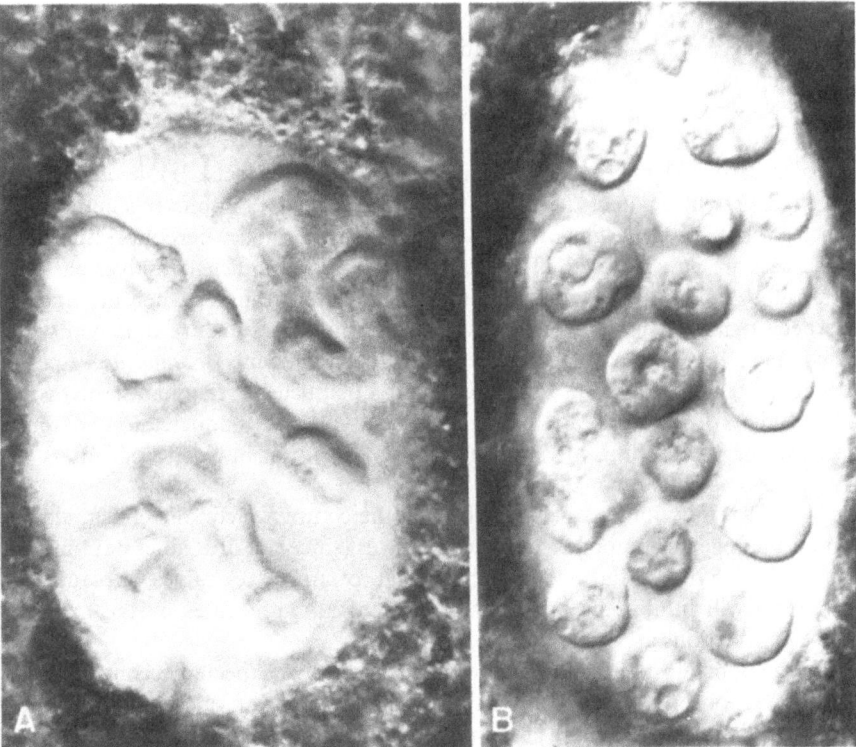

Fig. 66 A, B. Living nuclei of two species of *Acetabularia* photographed in situ with interference optics. **A** *A. mediterranea:* sausage-shaped nucleolar aggregates. **B** *A. cliftonii:* separate nucleolar units (Spring et al. 1978)

tive nucleus is, like the oocyte germinal vesicle, in meiotic prophase. In both cases, the nucleoli are RNA-rich and well-developed; the chromosomes are in an extended lampbrush state and the nuclear sap is abundant. However, the life cycles diverge after meiosis: in *Acetabularia,* the repeated cell divisions of the secondary nuclei precede the formation of gametes and zygotes; in eggs, cleavage follows fertilization.

1.3 Biochemical Studies

1.3.1 Morphogenetic Substances and mRNAs: Experiments with Inhibitors

The main problem is, of course, the *chemical nature of the morphogenetic substances.* There are many good reasons for believing that since they are synthesized in the nucleus and carry the information necessary for the production of the many proteins which are certainly needed for cap formation, the morphogenetic substances are a *family of stable mRNA molecules*. No experiments contradicting this hypothesis have so far been published and there is, as we shall see, ample circumstantial evidence in its favor. However, direct proof, such as isolation of the mRNAs produced by the nucleus and demonstration that they have biological activity, is still missing.

In favor of the mRNA hypothesis are a number of facts. *UV-irradiation*, at wavelengths which affect the integrity of nucleic acids, inhibits regeneration in anucleate halves, in particular when the apex of the stalk, where morphogenetic substances are accumulated, is irradiated. Irradiation of nucleate halves has only a transitory effect. Regeneration is stopped and then resumes when the nucleus has been damaged in a reversible way only. Similar results are obtained when living fragments of the algae are treated with the enzyme *ribonuclease*. Loss of regeneration is irreversible after treatment of anucleate fragments, reversible after action of ribonuclease on the nucleate ones. Ribonuclease apparently destroys the pre-existing mRNAs; new mRNAs cannot be produced unless the nucleus is present.

The mRNA hypothesis predicts that, on the contrary, actinomycin should not inhibit the regeneration of the anucleate halves (since the drug does not bind the pre-existing RNAs), but would arrest that of nucleate halves (because actinomycin will inhibit the synthesis of new mRNA molecules by the nucleus). These expectations are fulfilled by experiments.

In actinomycin-treated anucleate fragments, an unexpected secondary effect is, however, observed after a few days. While the initiation of cap formation was as good as in the controls, the growth of the anucleate caps is definitely depressed. The probable explanation of this secondary effect is the following: the *chloroplasts,* which exist in very large numbers in the algae, contain their own DNA. Actinomycin binds to chloroplastic DNA and this binding results in alterations of chloroplast activity which lead to a slowing down of the outgrowth of the cap.

Autoradiography brings further evidence for the view that the morphogenetic substances are stable mRNAs. If a normal alga is treated for a short time with labelled uridine, this precursor is very quickly incorporated into nuclear RNA. If the algae are cultivated, after this uridine "pulse", in normal seawater, one can follow the migration of the RNAs which have been synthesized in the nucleus. They first move into the cytoplasm, then accumulate at the apex of the stalk (where, as we know, the morphogenetic substances are also concentrated). We shall soon see that more recent biochemical evidence also support the stable mRNA hypothesis.

1.3.2 Energy Production

Oxygen consumption and *photosynthesis* remain normal in anucleate fragments for a few days. Afterward, the rate of photosynthesis somewhat decreases, as well as the ATP content, in the anucleate halves.

Of particular interest is the existence of a *circadian rhythm* in photosynthetic activity. If the algae are submitted to alternate periods of light and darkness, each period lasting about 12 h, they "remember" this photosynthetic rhythm for a certain time when they are cultivated in continuous light. Anucleate fragments retain their circadian rhythm of photosynthesis for many weeks. If however, nucleate fragments with a given rhythm are grafted on anucleate halves which have a different periodicity in their rhythm, the nucleus imposes its rhythm on the chloroplasts of the anucleate halves. It is even possible to combine, in such grafting experiments, nucleate halves which have lost their rhythmicity with anucleate fragments which possess a normal rhythm or vice versa. Only when the nucleate half has retained its rhythm will the graft be the site of a circadian rhythm.

This, as well as other experiments made with inhibitors of RNA and protein synthesis, has shown that the circadian rhythm of photosynthesis is not completely autonomous in anucleate halves. A nuclear control, probably exerted by mRNAs, superimposes itself on cytoplasmic rhythmicity and controls it.

Similar results and conclusions have been obtained for several other circadian rhythms found in *Acetabularia*. They deal with the shape of the chloroplasts, the RNA, polysaccharide and ATP content, etc.

1.3.3 Protein Synthesis

The protein content of both nucleate and anucleate halves approximately doubles during the 2 weeks which follow the sectioning of the algae. Since many of the newly synthesized proteins are specific *enzymes* and since we know that enzyme synthesis is controlled by the genes, it follows that the information which emerges from the genes must be stored in the cytoplasm of the anucleate fragments. The only explanation that molecular biology can offer for such a paradox is the intervention of stable mRNA molecules, which have been synthesized in the nucleus and stored in the cytoplasm.

Certain enzymes, for instance, the phosphatases and the enzymes involved in the production of the cell walls, markedly increase in activity when the caps form. This increase occurs in anucleate as well as in nucleate fragments. Some of these enzymes are distributed along the apico-basal gradient as morphogenetic substances, a fact which suggests that the corresponding mRNAs might have the same spatial distribution.

Regeneration, in anucleate fragments, is quickly and irreversibly blocked by *puromycin* and *cycloheximide,* the two classical inhibitors of cytoplasmic protein synthesis. On the other hand, *chloramphenicol* and *tetracycline,* which inhibit chloroplastic and mitochondrial protein synthesis, have very little effect on regeneration. Since the effects of puromycin and cycloheximide on regeneration are reversible in the case of nucleate fragments, one can safely conclude that proteins synthesized on cytoplasmic polysomes, under the direction of mRNAs of nuclear origin, play a more important role in morphogeneis than the newly synthesized chloroplastic proteins.

The chlorophasts contain their own enzymes and it would seem likely, at first sight, that this synthesis is controlled by chloroplastic rather than nuclear DNA. Experiments by Schweiger, in which nucleate halves of A. *mediterranea* were grafted into anucleate halves of *A. crenulata* and vice vers, have shown that this is not true for the lactic and malic dehydrogenases. The enzymes present in the chloroplasts of the "hybrid" progressively change and are replaced by others, which are characteristic of the species which provided the nucleus. The same nuclear control has been found for the synthesis of the insoluble proteins which form the chloroplastic membrane and of the chloroplastic ribosomal proteins.

Inhibitors which discriminate between chloroplastic and cytoplasmic protein synthesis have been very useful for the analysis of enzyme synthesis in *Acetabularia*. As we have seen, puromycin and cycloheximide inhibit the functioning of the cytoplasmic 80S ribosomes, while chloramphenicol blocks protein synthesis in the chloroplastic 70S ribosomes and in the still smaller mitochondrial ribosomes. Use of these inhibitors has disclosed an unexpected fact: one would have expected that the activity of the enzymes involved in DNA synthesis increases when the vegetative nucleus breaks down and the daughter nuclei multiply at a fast rate. This is indeed what happens for thymidine kinase and other enzymes important for DNA synthesis; but a surprise came when it was found that these enzymes are encoded by chloroplastic DNA, not by nuclear genes.

Subtle interactions between the nuclear and chloroplastic genomes thus take place during the life cycle of the alga, in particular at the time of cap formation. In order to throw some light on these interactions, a direct approach has been taken recently by H. G. Schweiger and colleagues. They injected mRNAs into the algae and studied their in vivo translation on 80S cytoplasmic ribosomes. The main conclusion was that during the *Acetabularia* life cycle, translation of some proteins is not regulated, i.e. it proceeds for months at a constant rate; but translation of other proteins is turned off at a given stage. In contrast, at the same stage, translation of still another group of proteins is enhanced. Continuation of these experiments should throw more light on the complex pattern of protein synthesis regulation during growth and morphogenesis in *Acetabularia*.

1.3.4 RNA Synthesis

The total RNA content, like the protein content, doubles within 2 weeks when caps are formed in either nucleate or anucleate fragments. This large increase is mainly due to the synthesis of *chloroplastic* RNAs. Only recent work, in the laboratories of H.G. Schweiger and W. Franke, has allowed precise studies on RNA production by the nucleus.

Amplification of the *ribosomal* (nucleolar) genes occurs, as in amphibian oocytes, in the *Acetabularia* vegetative nucleus. Electron microscopy of spread nucleoli and estimations of their DNA content have shown that there are about 4000 copies of the 28S and 18S rRNA genes in the vegetative nucleus. As in the oocyte nucleus, this number of copies does not increase markedly during the growth period of the vegetative nucleus, thus, amplification apparently takes place soon after zygote formation.

The general organization of the ribosomal genes is very similar in the nucleolar organizer of *Acetabularia* and *Xenopus* oocytes (see Figs. 27 and 28). Tandem repeated transcription units are separated by untranscribed spacers of variable lengths. However, these spacers are, on the average, smaller in *Acetabularia* than in *Xenopus* and the size of the nucleolar rRNA precursor, in the alga, is not much larger than that of the final products (28 S + 18 S rRNAs). The rate of rRNA synthesis by the vegetative nucleus is very high (4×10^7 nucleotides/per second per nucleus); it is increased 20 times a few days after removal of part of the stalk. The ribosomal RNAs are quickly transferred from the nucleus to the cytoplasm, where they display a great stability (half-life of 80 days); they move from the rhizoid to the apex of the stalk with a speed of 2–4 mm day^{-1}. As one can see, the *Acetabularia* nucleus is, like the oocyte nucleus, a machine for the large production of cytoplasmic ribosomes.

Fortunately, the chloroplastic mRNAs have no polyadenylic "tail" (see Chap. III) and this has given the possibility of selectively "catching" the *messengers* produced by the vegetative nucleus. As expected, these mRNAs of nuclear origin are synthesized only by nucleate fragments of the alga and are heterogeneous: their molecular weights range between 0.5 and 3×10^6. They migrate from the nucleus to the apex of the stalk independently of the rRNAs at a speed of 5 mm day^{-1} (the growth rate of the stalk itself is only 1–3 mm day^{-1}), thus independently from the ribosomal RNAs.

The increase in the rate of mRNA synthesis (two- to threefold) after cutting the alga into two halves, is much smaller than for the rRNAs. These data suggest that transcription of chromosomal DNA plays only a limited role in the control of genetic activity and morphogenesis in *Acetabularia*. The major role is played (as in maturing *Xenopus* oocytes and fertilized sea urchin eggs) by the selective translation of mRNAs of nuclear origin stored at the apex of the alga.

Little is known, unfortunately, about the nature of the factors which regulate the translation of the mRNAs accumulated at the apex of the stalk. Inhibitors of *proteolytic enzymes* prevent cap formation in both nucleate and anucleate fragments: it might be that protease activity is required for the "unmasking" of the mRNAs of nuclear origin which are linked to proteins in the form of ribonucleoprotein particles. Another regulatory factor is the *polyamine* content: inhibitors of ornithine decarboxylase (ODC) prevent cap formation in both kinds of fragments. They also prevent the formation of cysts, if they are added to whole algae after the breakdown of the vegetative nucleus. Since the ODC inhibitors decrease the polyamine content of treated cells, it can be concluded that the putrescine, spermidine and spermine levels control both RNA and DNA synthesis in the alga. In polyamine-depleted algae, the decrease in RNA synthesis is followed by inhibition of growth and morphogenesis; the slowing down of DNA synthesis in the secondary nuclei prevents normal cyst formation.

These experiments with ODC and proteolytic enzyme inhibitors have allowed an estimation of the stability of the morphogenetic substances in anucleate fragments of *Acetabularia*. If such fragments are treated with the inhibitors during different lengths of time and then cultured in normal medium, reversibility becomes partial after a 10-day treatment and completely disappears after 2–3 weeks. This is in good agreement with the estimated half-life of the *Acetabularia* messengers (about 10 days) and provides further evidence for the view that Häm-merling's morphogenetic substances are a mixture of mRNAs synthesized by the nucleus and stored in the apical cytoplasm.

1.3.5 DNA Synthesis. Relative Autonomy of the Chloroplasts

As we have seen, there is no replication of nuclear DNA during the whole veg-etative life of the alga (which lasts several months). Only when the cap has reached its maximal size does the big vegetative nucleus break down and give rise to daughter nuclei which replicate their DNA, divide and are transferred passively to the cap by protoplasmic streaming.

In contrast, *chloroplastic* and *mitochondrial* DNAs are very actively synthe-sized by both nucleate and anucleate fragments of *Acetabularia*. The total DNA content increases two to three times in both kinds of fragments during the 10 days which follow the sectioning. This came as a big surprise 20 years age, since the very existence of DNA in chloroplasts seemed very doubtful at the time. Now we know that chloroplastic and nuclear DNAs are very different molecules. In *Acetabularia*, chloroplastic DNA is a mixture of large circular and long linear molecules. In addition, small circular molecules, corresponding to replicating chloroplastic DNA, have been observed under the electron microscope (review by Lüttke and Bonotto). The use of highly sensitive and very specific stains for DNA allows us to see chloroplastic DNA with a fluorescence microscope, which is possible because the DNA content of the *Acetabularia* chloroplasts is excep-tionally high as compared to that of other unicellular algae. Fluorescence mi-croscopy shows that the DNA content of all *Acetabularia* chloroplasts is not the same, i.e. the small, young chloroplasts, which are accumulated at the tip of growing algae, contain much more DNA than the large chloroplasts which sur-round the nucleus. In fact, many of these basal chloroplasts, which are heavily loaded with carbohydrate reserves, have probably no DNA at all. There is thus a progressive loss of chloroplastic DNA along the apico-basal gradient.

Chloropastic DNA can, in *Acetabularia* as elsewhere, be transcribed and translated. If they are given light, isolated chloroplasts synthesize all kinds of RNAs and a number of proteins (in particular, proteins involved in the formation of chloroplastic membranes and lamellae). However, as we have seen, the syn-thesis of enzymes such as lactic and malic dehydrogenases and that of the chloro-plastic ribosomal proteins requires the co-operation of the nucleus: the *autonomy* of the chloroplasts toward the nucleus is thus imperfect. Counts of the chloro-plasts, in nucleate and anucleate fragments of the alga, have shown that they in-crease in number even in the absence of the nucleus. However, chloroplast mul-tiplication is more active in the nucleate than in the anucleate halves, where many chloroplasts are large and have a dumbbell shape. It is thus probable that cyto-plasmic factors, which are synthesized under nuclear control, are required for fission of the chloroplasts and that they are missing in the anucleate halves. Since even an apparently specific chloroplastic function such as the circadian rhythm of photosynthesis is regulated on cytoplasmic (80S) ribosomes, we must again conclude that the chloroplasts are not fully autonomous toward the nucleus. In whole algae, the chloroplastic and nuclear genomes co-operate in order to specify the protein composition of the chloroplasts and to allow their multiplication.

1.3.6 Complexity of Nucleocytoplasmic Interactions

It is clear that the nucleus provides, in *Acetabularia,* the various mRNAs which are required for morphogenesis (cap formation), i.e. the nucleus is the source of the morphogenetic substances and therefore exerts a very important *positive control* on the cytoplasm. It should be emphasized that although anucleate fragments can utilize their stored mRNAs of nuclear origin and form full-sized caps, these caps always remain sterile: neither cysts, nor gametes form in anucleate caps which are therefore useless for reproduction purposes.

But it would be a mistake to believe that the vegetative nucleus of *Acetabularia* exerts only positive effects on the cytoplasm. Already 50 years ago, Hämmerling demonstrated that anucleate fragments of *Acetabularia* form caps faster than control whole algae. Similarly, we found that protein synthesis is accelerated when the nucleus has been removed. There is no doubt that the nucleus exerts *negative,* as well as positive controls over the cytoplasm.

In addition, the *cytoplasm* controls nuclear activity. If the photosynthetic function of the chloroplasts is suppressed by culture in darkness, the nucleus shrinks. The nucleolus undergoes vacuolization and stops synthesizing RNA. Thus, energy production in the cytoplasm is absolutely required for the maintenance of nuclear morphology and synthetic activity. If, as shown long ago by Hämmerling, one cuts repeatedly an alga into halves when it begins to form a cap, the breakdown of the large vegetative nucleus is inhibited, thus there will be no formation of daughter nuclei, no nuclear DNA synthesis. Repeated amputation of the cap thus leads to immortality. As in amphibian oocytes undergoing maturation, the choice between DNA transcription and replication is decided by cytoplasmic factors. This conclusion is reinforced by more recent experiments in which an old vegetative nucleus was introduced into the cytoplasm of a young alga; this nucleus did not break down and one can speak of "rejuvenation" of the old nucleus by the surrounding young cytoplasm.

As we shall now see, although a giant green alga seems very different from an amphibian or sea urchin egg, there are remarkable similarities in the nucleocytoplasmic interactions in all cases. We are again faced with the paradox if unity and diversity in living organisms.

2 Biochemistry of Anucleate Fragments of Eggs

2.1 Amphibian Eggs

We have seen in Chap. IV that the huge germinal vesicle plays only a passive role in the biochemical changes (increase in protein synthesis with preferential histone synthesis, stimulation of protein phosphorylation, production of the maturation promoting factor) which take place when maturation is induced in full-grown *Xenopus* oocytes. All these changes occur even if the oocytes have been enucleated prior to progesterone addition. It has also been recently shown that a "heat shock" (heating for a few minutes above 30 °C) induces the synthesis of a few new proteins in *Xenopus* oocytes. These heat shock proteins are the same whether the oocyte possesses its nucleus or not. Enucleated *Xenopus* oocytes contain the cytoplasmic factors required for "reprogramming" injected nuclei. As we have seen in Chap. IV, the cytoplasm of these oocytes contains factors which inactivate certain genes and reactivate others. The outcome of the experiments in which protein or 5 S RNA synthesis has been followed after injection of adult nuclei is the same whether the oocytes have a nucleus or not. Finally, enucleated oocytes undergo the same permeability changes and the same cortical reaction after parthogenetic activation as intact progesterone-treated oocytes.

However, we have seen that the mixing of the nuclear sap and the cytoplasm is required for swelling of injected nuclei, condensation of chromatin and aster formation. It has been reported that enucleated amphibian oocytes undergo repeated cleavages after injection of sea urchin mitotic apparatuses: however, it is possible that in such experiments, sea urchin chromosomes take part in the cleavages.

Cleavage in the absence of the nucleus has been obtained repeatedly in amphibian eggs. In these experiments, the eggs were fertilized and the egg and sperm nuclei were destroyed by pricking or irradiation. In general, partial blastulae, in which only the animal pole has cleaved, are obtained; gastrulation never takes place. The morphogenetic potencies of these *"achromosomal"* blastulae have been tested by grafting nonnucleated "cells" from arrested blastulae onto inducing sites of normal gastrulae. The grafts survived up to 4 days, but showed no sign of differentiation: nonnucleate ectoderm cells are thus totally unable to respond to the inducing stimulus of a normal organizer. These experiments show that the morphogenetic potencies of enucleated amphibian eggs are much lower than those of anucleate fragments of *Acetabularia*: they are limited to the capacity to divide repeatedly. We shall now see that the same is true for anucleate fragments of sea urchin eggs, which have been better studied from the biochemical viewpoint.

2.2 Sea Urchin Eggs

2.2.1 Biological Observations

The only materials that can be obtained in sufficient quantities for easy biochemical analysis are the anucleate fragments of sea urchin eggs. As already mentioned and as shown in Fig. 37, the unfertilized eggs of *Arbacia* can be cut into two halves by centrifugation in a density gradient. The heavy red pigment accumulates to the centripetal end of the egg, which successively elongates, takes a dumbbell shape and separates into two halves. The light, transparent half contains the nucleus, fat droplets, mitochondria and a clear, ribosome-rich layer, the hyaloplasm. The red heavy half is anucleate; it contains the red pigment granules and most of the yolk. Electron microscopy shows that the light halves also possess some of the yolk platelets and that mitochondria, as well as lipid droplets, are present in both halves. All kinds of egg inclusions are therefore presen in the two halves, but they are in different proportions; the only specific cell organelles are the nucleus, which is always in the light half, and the pigment granules, which are found only in the heavy halves.

Both kinds of fragments can be fertilized or parthenogenetically activated by the classical method of Loeb (hypertonic seawater and butyric acid treatments). Nucleate halves develop very well after *fertilization* and give rise to normal plutei. The development of fertilized anucleate halves is less normal and plutei are not easy to obtain. This rather poor development is probably not so much due to haploidy as to the excess of yolk, which makes cleavage difficult. *Parthenogenetic* development of the nucleate halves is almost as good as that of normal eggs. But the development of the anucleate halves, after parthenogenetic treatments, is completely abortive. Fig. 67 shows that one can, at best, hope to get an irregular fragmentation of the eggs rather than a true cleavage. Asters can repeatedly form in the cytoplasm of the activated anucleate halves and irregular furrows can cleave the egg to form a kind of morula. But the typical cleavage pattern (Fig. 13) of the sea urchin eggs, with micromeres, mesomeres and macromeres is always missing. The blastula-like larvae neither hatch nor form cilia. As one can see, there is no true morphogenesis and we are very far indeed

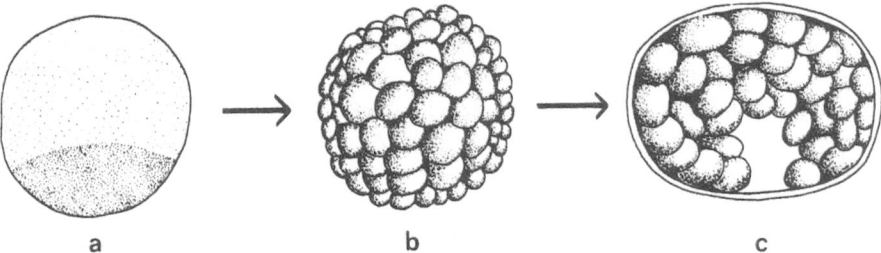

| a | b | c |

Fig. 67 a Anucleate half of a sea urchin egg. **b** Anucleate morula after 13 h. **c** Anucleate blastula after 3 days (E. B. Harvey)

from the formation of the elegant, species-specific cap of *Acetabularia*. Only the formation of asters and furrows displays some autonomy toward the nucleus in activated anucleate sea urchin eggs.

The only other anucleate materials that have been obtained from eggs and have been the subject of rather limited biochemical studies are the *polar lobes,* which can be separated at the "trefoil" stage from the eggs of the mollusks *Mytilus* and *Ilyanassa*. They never form asters and do not cleave; all one can see are localized contractions of the cortex, which confirm that contractility of the cortical proteins displays some autonomy toward the nucleus.

2.2.2 Biochemical Studies

These results can be presented briefly, since the essentials have already been related in preceding chapters. We shall first consider energy production, then synthesis of macromolecules.

The *respiration* of nucleate and anucleate halves of unfertilized sea urchin eggs is, like that of the whole eggs, very low. If the fragments are fertilized, or if they undergo parthenogenetic activation, their oxygen consumption quickly increases. The respiration of the anucleate halves becomes even higher than that of the nucleate ones. This proves that the part played by the female pronucleus, in cellular oxidations, is negligible. This conclusions agrees with the already mentioned finding that the oxygen consumption of the isolated germinal vesicle of frog oocytes is hardly measurable. In fact, enucleated and normal frog oocytes have about the same respiratory rate. The higher oxygen consumption of the anucleate halves is probably related to the fact that, as shown by electron microscopy, the nucleate and anucleate halves contain, after their separation by centrifugation, two different populations of mitochondria. They differ in size, structure and probably in respiratory enzyme content. In fact, the anucleate, heavy halves contain a larger proportion of the reducing mitochondrial respiratory enzymes (dehydrogenases) than their nucleate counterparts.

Isolated polar lobes of the mollusk *Mytilus* have a low respiratory rate, which does not increase with time, in contrast with the blastomeres where the nuclei multiply at a fast pace.

Coming now to the synthesis of macromolecules, we have already seen that *protein synthesis* is independent of the presence of the nucleus in sea urchin eggs. Activation of anucleate fragments is followed by a strong stimulation of protein synthesis and is even stronger in activated anucleate halves than in nucleate ones. This stimulation is accompanied by an increase in the number of polysomes in anucleate as well as in nucleate halves; these polysomes result from the binding of "unmasked" maternal mRNA molecules to the pre-existing ribosomes.

A whole spectrum of proteins are synthesized after activation in both nucleate and anucleate halves, but it is not yet known whether the neosynthesized polypeptides are identical or slightly different in the two halves. Among the stored mRNAs, which are translated by both kinds of activated fragments, those coding for *histones* and for *tubulin* deserve special interest for the following reasons. In anucleate halves, there is of course no replication of nuclear DNA and no assembly of DNA in nucleosomes; yet histones are synthesized, apparently uselessly. This definitely proves that there is not necessarily a close link between histone production and nuclear DNA synthesis. Synthesis of tubulin in activated anucleate halves also raises an interesting question: Is translation of tubulin mRNA involved in the formation of asters in activated anucleate fragments? It is well known that unfertilized sea urchin eggs have a large pool of unpolymerized tubulin molecules. The formation of asters and spindle, in the fertilized sea urchin eggs, results from their polymerization around microtubule organizing centers (MTOC), such as the kinetochores of the chromosomes and the centrioles. If anucleate halves are treated with heavy water (D_2O), many cytasters appear, which are centered around small centrioles or precentrioles which have apparently been assembled de novo. It is not known whether translation of tubulin mRNA is involved in the formation of asters in activated anucleate sea urchin egg fragments. But it is a striking fact that, despite tubulin synthesis, anucleate activated halves never form cilia; anucleate amphibian blastulae also lack cilia. Something necessary for the assembly of cilia is thus missing in nonnucleate cytoplasm. Whatever this might be, it seems clear that its production, in blastulae, requires transcription of nuclear genes. Anucleate parthenogenetically activated sea urchin fragments are also unable to hatch: they apparently cannot synthesize the "hatching enzyme". It would be interesting to know whether they possess a maternal mRNA encoding this proteolytic enzyme.

The *distribution* of the maternal mRNAs in the nucleate and anucleate halves of sea urchin eggs (obtained by high speed centrifugation) has been recently studied. It was found that total RNA, the bulk of the poly A containing mRNAs, actin and tubulin mRNAs are accumulated in the heavy, anucleate halves. This might explain why protein synthesis is stronger, after parthenogenetic stimulation, in anucleate than in nucleate halves. In contrast, histone mRNA is accumulated in the nucleate halves (even in uncentrifuged eggs cut into halves with a glass needle). However, isolated nuclei contain little histone mRNA. The distribution of the various mRNAs is thus heterogeneous in sea urchin eggs, which is probably due to the organization of the cytoplasm in gradients.

If we compare amphibian and sea urchin eggs with *Acetabularia,* a striking fact emerges: although all anucleate fragments possess a large store of mRNAs, only the alga is capable of true morphogenesis. The reasons why, in eggs, the maternal mRNA store (theoretically capable, as we have seen, to synthesize more than 10,000 different proteins) does not allow even such early events as hatching and ciliation, are not known. It might be that the mRNAs are less stable in anucleate than in nucleate fragments. However, it is known that polyadenylation of the same mRNAs occurs in both kinds of fragments and that the enzyme responsible for polyadenylation is accumulated in the heavy anucleate halves. As we know, one of the functions of poly A addition is to increase mRNA stability in the cytoplasm. Transcription of nuclear genes during and after cleavage is thus an absolute prerequisite for further development. In molecular terms, the maternal mRNA store corresponds to *preformation,* the neosynthesized mRNAs to *epigenesis:* that development is an epigenetic process is, of course no surprise for embryologists. Development stops, in enucleated eggs, at an early cleavage stage probably because they are unable to synthesize the mRNAs and the specific proteins which are required for further developmental events.

Both nucleate and anucleate fragments of sea urchin eggs possess a large population of *mitochondria.* Since they contain DNA, the DNA content of nucleate and anucleate halves is about the same (the haploid female pronucleus represents only a small proportion of total DNA). After activation, both kinds of fragments synthesize small amounts of mitochondrial RNAs (in particular, mitochondrial ribosomal RNAs). The mitochondria of activated anucleate fragments synthesize as RNA species which can diffuse into the cytoplasm, but, since it is unable to bind to the cytoplasmic ribosomes, it cannot play an important role in the strong stimulation of protein synthesis which follows activation of anucleate halves.

An unexpected fact has recently come to light: *mitochondrial* DNA, RNA and protein syntheses are more strongly stimulated by parthenogenetic activation in anucleate than in nucleate halves of sea urchin eggs. It seems that removal of the nucleus induces a multiplication of the mitochondria, as well as an increase in mitochondrial DNA transcription and translation. If so, the often neglected female pronucleus would exert a *negative* control on the mitochondrial genome. This is reminiscent of the negative control exerted by the *Acetabularia* nucleus on the chloroplastic genome. In this alga, removal of the nucleus speeds up DNA, RNA and protein synthesis and we have seen that the chloroplasts are largely responsible for this increase in macromolecule synthesis.

As already shown in Fig. 12, removal of the *polar lobe* at the trefoil stage results, in the mollusk *Ilyanassa,* in developmental abnormalities. Isolated polar lobes synthesize many proteins and thus contain, like sea urchin eggs, a store of maternal mRNAs. Removal of the polar lobe does not affect RNA and protein synthesis before relatively late stages of embryogenesis: no changes can be detected before the first appearance of developmental abnormalities. There is no evidence so far that the polar lobe synthesizes specific proteins and we have therefore no indication for a selective localization of certain translatable mRNAs in this region of the *Ilyanassa* egg. Curiously, the pattern of protein synthesis changes when isolated polar lobes are allowed to age. These changes are the same as those observed in intact embryos and should result from a selective translation of the maternal mRNAs (which is responsible for 98% of protein synthesis during early embryogenesis in *Ilyanassa*). A recent study concludes that the cytoplasm of the polar lobe contains both repressors of translation and activators for unmasking maternal mRNAs: if so, both positive and negative controls of mRNA translation would operate in pure, nonnucleate cytoplasm.

3 The Importance of the Nucleus for Embryonic Development

3.1 General Background

Overwhelming evidence demonstrates that integrity of the nucleus is required for development beyond the blastula stage: haploidy, aneuploidy (unequal distribution of the genetic material in the blastomeres), in some cases hybridization. Many mutations are sooner or later *lethal* for the embryo. We shall concentrate on the cases in which lethality is an early event (arrest at gastrulation or neurulation) because they demonstrate that even primary morphogenesis is under nuclear control.

Haploidy can be obtained, in amphibian eggs, by a variety of means (pricking with a needle, fertilization with UV-irradiated sperm, removal of the maturation spindle just after fertilization). Whether the haploid embryos arise from the egg nucleus (*gynogenesis*) or the sperm nucleus (*androgenesis*), the result is the same. After normal cleavage, gastrulation is slowed down and further development leads to the already mentioned *haploid syndrome,* which is characterized by microcephaly, deficiency in blood circulation, reduction of the gills, oedema, ascites and finally death at an early larval stage. Lethality seems to be due to haploidy per se, and not to lethal genes which would not be balanced by their normal alleles in the haploid condition. If, as may happen accidentally, the number of chromosomes doubles in a parthenogenetic egg before it undergoes cleavage, homozygous diploids are produced. Although they contain two copies of the lethal genes, they can reach the adult stage. This is apparently not true for the mouse in which haploid eggs fail to develop beyond the end of cleavage and parthenogenetic diploids are lethal.

In *polyspermic* eggs, only one of the nuclei of the supernumerary spermatozoa fuses with the egg pronucleus; the other nuclei remain haploid and may take part in the development. The resulting embryos are a mosaic of diploid and haploid nuclei; although part of the embryo is diploid, death occurs still earlier than in haploids. In mammals, dispermy, which takes place spontaneously in many species, including man, gives rise to *triploid* embryos. Triploids may also arise from fertilization of a diploid egg. It is lethal during the postimplantation development; only very few triploid embryos have been reported to have developed until term. Spontaneously and cytochalasin B-induced tetraploid mammalian embryos display the same lethality. Interestingly, mosaic embryos constructed by aggregation of diploid and triploid or tetraploid morulae are also lethal.

Aneuploidy (unbalanced chromosome complement) always has serious consequences: if, only one or two chromosomes are lost or added to the normal complement, viability is greatly reduced and morphogenesis is abnormal. If chromosome unbalance is very marked, as a result of multipolar mitoses during early cleavage, for instance, development stops at the blastula stage. Strong aneuploidy is thus as lethal as the complete absence of a nucleus.

A great deal of interest has been recently devoted to aneuploidy in mammals because it is the major cause of prenatal death and abortion or of severe postnatal congenital defects such as the human Down's syndrome. Monosomic mouse embryos (39 chromosomes instead of 40) die before or at implantation, while trisomic (41 chromosomes) embryos survive until at least day 10 of pregnancy. It is possible to rescue cells from such aneuploids by aggregating aneuploid morulae with normal diploid morulae. This strongly suggests that the chromosome unbalance is not lethal to the cells, but is specifically responsible for the impairment of morphogenetic events.

In the following, we shall deal mainly with lethal *hybrids,* because they have been better studied from the biochemical viewpoint than haploids, aneuploids and polyspermic embryos.

3.2 Lethal Hybrids

3.2.1 Biological Observations

Various things can happen when eggs are fertilized with sperm from a different species. The spermatozoon, in many cases, does not even touch the surface of the egg and nothing happens at all. In other cases, the spermatozoon comes in contact with the egg membrane and elicits the activation reaction, but it does not penetrate into the egg and does not even act as a parthenogenetic agent. In the remaining cases, the spermatozoon penetrates into the egg and amphimixy follows. But, in many instances, the paternal chromosomes, which have been introduced into the egg by the spermatozoon, are eliminated into the cytoplasm during cleavage; the eliminated chromosomes soon degenerate and development is of the haploid, gynogenetic type (*pseudohybrids*). In true hybrids, the paternal chromosomes are not eliminated; development is normal, for a longer or shorter time; then the hybrid embryos develop abnormalities and ultimately die. They are the so-called *lethal hybrids*. Finally, when the chromosomes of the two parental species are sufficiently similar from the genetic viewpoint, the hybrids are *viable;* they reach the adult stage, but they are not always able to reproduce. Sterility is frequent, because the homologous chromosomes cannot undergo perfect pairing during meiosis.

Early lethal hybrids will be the main subject of this chapter, because they have a peculiar interest for molecular embryology. When eggs of the common species of sea urchins or frogs are fertilized with sperm from other common species, development is often arrested at the late blastula or *early gastrula* stage. This is the stage at which, as we have seen, major biochemical changes occur in the embryo; the most conspicuous of them is the resumption of rRNA synthesis. Why the hybrid embryos are unable to go through this critical stage of gastrulation is the question that a number of biochemical studies have tried to answer.

Before we discuss the biochemical studies on early lethal embryos, a last important biological fact must be mentioned. If a piece of an amphibian lethal hybrid, which is arrested in development at the early gastrula stage, is grafted into a normal host, a "revitalization" usually follows. If, for instance, the dorsal lip of a blocked hybrid gastrula is grafted into a normal recipient gastrula, the grafted dorsal lip differentiates into chorda and induces, in the host, a secondary nervous system. This revitalization is not species-specific. The same result (chorda differentiation in the graft and neural induction in the host) is obtained when a dorsal lip taken from an hybrid between two species of frogs is grafted in a urodele (triton, axolotl) gastrula.

Revitalization has been explained by supposing that unspecific substances, which cannot be synthesized by the hybrid cells, are given by the host cells to the graft. But it has recently been shown that for certain lethal hybrids at least, it is sufficient to dissociate the cells of the blocked embryo and to cultivate them in salt solution to obtain extensive differentiation of the hybrid cells. The "conditioned media" prepared from such hybrid cells are very toxic for normal embryos. Therefore, it is possible that lethality is due, at least in part, to the production and retention of toxic substances (repressors) produced by metabolism.

Nuclear transplantation has been used in order to find out whether the block at gastrulation in the combination *R. pipiens* ♀ × *R. catesbeiana* ♂ is due to irreversible changes in the nuclei. Transfer of nuclei taken from arrested hybrid gastrulae into recipient enucleated *R. pipiens* unfertilized eggs showed that nuclei taken from hybrid embryos that have just stopped developing allow normal cleavage; but development stops abruptly at gastrulation (as in the hybrids themselves). If nuclei are taken 10 to 20 h after gastrulation arrest and then transplanted into enucleated unfertilized eggs, cleavage is no longer normal. There are thus no irreversible changes in the nuclei of the lethal hybrids so long as they do not undergo cytologically visible alterations.

3.2.2 Biochemical Studies on Lethal Hybrids

3.2.2.1 Respiration. Most of the biochemical studies made on lethal hybrids between sea urchins have been done with the combination *Paracentrotus* ♀ × *Arbacia* ♂. These hybrids cleave normally, hatch and die during gastrulation.

Paracentrotus (P) eggs have a higher oxygen consumption than those of *Arbacia* (A). The respiration of the P × A hybrid is first similar to that of the maternal species (P). Later on, the respiratory rate decreases in the hybrids, so that intermediary values between those typical for P and A are obtained. This reduction of the respiratory rate in the P × A hybrids suggests that a negative control, due to the paternal A genes, becomes effective a few hours before the arrest of development.

More data are available for hybrids among amphibians. The best studied combination is one among two American species of frogs: *Rana pipiens* ♀ × *Rana sylvatica* ♂, where development stops at gastrulation. Respiration shows a normal increase during cleavage, then remains absolutely constant until the death of the embryo several days later. This decrease in the respiratory rate affects all parts of the embryo, as shown by measurements of the oxygen consumption made on dissected hybrid gastrulae. The decrease in respiration rate is the same in the ectoblast, the mesoblast and the entoblast. The reduced respiratory rate of the hybrids is linked, as one would expect, to a diminution of carbohydrate metabolism. Glycogenolysis is slowed down, glycolysis is decreased and the ATP content drops more quickly under anaerobic conditions in the hybrid gastrulae than in the controls after development has stopped.

These results cannot be generalized to all hybrid combinations between amphibians, however. The rule seems to be that as in sea urchin hybrids the oxygen consumption still increases, but at a slower rate than in the controls in arrested embryos.

3.2.2.2 Nucleic Acid and Protein Synthesis. We shall first consider the lethal hybrids of echinoderms, then those of amphibians.

Protein synthesis has been studied in several combinations of *sea urchins.* The "pool" of soluble amino acids and small peptides remains of the maternal type, even at a time when an effect of the paternal nucleus should become evident. In the *Paracentrotus* ♀ × *Arbacia* ♂ combination (P × A), paternal proteins (antigens) can be detected, by immunological methods, in the hybrid blastula. The amount of paternal protein synthesized is about half that which would have been produced by a diploid *Arbacia* (AA); it is identical with the quantity of *Arbacia* antigen which is produced by a haploid, parthenogenetic *Arbacia* embryo at the same stage. These antigens are not synthesized at all by anucleate fragments of *Arbacia*. It follows that the amount of antigen produced depends on the number of *Arbacia* genes present. It does not matter, in this respect, whether a haploid set of genes is placed in its own *Arbacia* cytoplasm or in the foreign *Paracentrotus* cytoplasm.

However, in other echinoderm combinations (*Strongylocentrotus* ♀ × *Dendraster* ♂) no paternal antigens can be detected in the hybrid embryos, which stop development at a later stage than in the P × A combination. In these hybrids, one enzyme has been characterized: it is the "hatching enzyme" which enables the blastula to dissolve the fertilization membrane and to swim freely in the sea. This hatching enzyme has been found in the hybrids to be of the purely maternal type. Taken together, these results suggest that in contrast to the P × A combination, the paternal chromosomes are partly or entirely inactivated in the hybrids which show a complete maternal dominance.

This suggestion has been reinforced by recent work on sea urchin hybrids which can reach the pluteus larval stage. Among many proteins synthesized by these hybrids, only a few are specified by the paternal genome. Analysis of mRNA production in these hybrids has further shown that the paternal mRNAs are strongly underrepresented; however, paternal histones and their mRNAs are detectable already during early cleavage. It has also been found that some of the messages coding for paternal proteins are not translated in the hybrids. Production of foreign paternal proteins in the hybrids is thus repressed at two levels: transcription of the paternal genes and translation of the mRNAs (except for the histone genes and their mRNAs).

The controls exerted by the egg cytoplasm on the activity of the paternal genes differ whether development of the hybrids stops at gastrulation or reaches the pluteus stage. A thorough study, with the now available methods, of the early lethal hybrids is needed before one can explain this difference. The present evidence suggests that arrest at gastrulation takes place in eggs in which the foreign paternal genes are not repressed by the egg cytoplasm and are thus expressed. Development until a more advanced larval stage would be possible when the paternal genes are repressed to a large extent during early development. In this case, development would be almost parthenogenetic as in gynogenetic pseudohybrids. However, in these hybrids the foreign paternal chromatin far from being eliminated, is repeatedly replicated.

Arrest of DNA synthesis is certainly *not* responsible for the block in development. DNA synthesis continues when the P × A embryos are blocked at the gastrula stage, but the rate of synthesis is slower than in the controls.

This probably results from a discrete elimination of paternal chromosomes during the development of the P × A embryos. Such an elimination also explains why the blocked P × A hybrids contain three times more maternal than paternal DNA, as shown by molecular hybridization.

The situation in *frog hybrids* is not fundamentally different, but a few peculiarities should first be mentioned. Hybridization modifies the properties of the plasma membrane; its permeability to amino acids and phosphate increases. Furthermore, if cells of a lethal hybrid gastrula are dissociated by removal of calcium and magnesium from the medium, they have great difficulties in reaggregating when these ions are added back to the medium.

These differences in reaggregation in hybrid and normal gastrula cells are probably due to differences in the surface glycoproteins; it is likely that changes in cell surface glycoproteins and cell adherence play an important role in the arrest of many hybrid combinations at gastrulation. Indeed, it is known that treatment of normal *Xenopus* morulae with lectins (which bind to the sugar residues of the surface glycoproteins) prevents their gastrulation.

Another characteristic of the lethal hybrids among frogs is that the *yolk* reserves are utilized at a much slower rate than in the controls. The enzymes present in the yolk platelets also fail to be synthesized or liberated at the normal rate. The reasons for this delay in vitellolysis are not known.

As in the lethal sea urchin hybrids, *paternal antigens* become detectable at the blastula stage and an additional basic protein is found in the hybrids.

The synthesis of several *enzymes* has been studied. Esterases remain of the maternal type in one lethal combination of frogs. In another, which is viable, the enzymes that catalyze the oxidative metabolism of the sugars have values that are intermediary between those found in the two parental species, but if the egg nucleus is removed before fertilization, in order to obtain a *merogone hybrid* (which is lethal at a later stage than gastrulation) the synthesis of these enzymes follows the paternal pattern.

As in the sea urchin hybrids, *DNA synthesis* does not stop, but only slows down after the arrest of development. The same is true for histone synthesis, which remains well coordinated with that of DNA. However, DNA and RNA syntheses stop before protein synthesis. When the hybrid is close to death and can no longer synthesize nucleic acids, it becomes comparable to the anucleate fragment of an egg, in which protein synthesis can continue for a long time because of the preformed maternal mRNAs and ribosomes.

A situation somewhat different from lethal hybridization has been achieved by Gurdon who injected, into an enucleated unfertilized egg of *Xenopus*, a diploid nucleus taken from a blastula of *Discoglossus*. In this case and in contrast with the lethal hybrids, a diploid foreign nucleus, but no maternal haploid nucleus, is present in the egg. The eggs that have received such a foreign *Discoglossus* nucleus stop development, as many lethal hybrids do, at the late blastula stage. RNA synthesis has been analyzed in these blocked blastulae. There is no rRNA synthesis and there is a decrease in mRNA and tRNA synthesis.

As one can see, much more work is needed before we can understand the molecular reasons for the lethality in both sea urchin and amphibian hybrids. If one can venture a guess, arrest of development, followed by cell death, might largely be the result of the accumulation of foreign, paternal proteins.

4 Biochemistry of Early Developmental Mutants

There is no doubt that molecular embryology would make tremendous progress if *mutants* deficient in a specific metabolic step were easily available. It is needless to recall that molecular biology would not be what it is today without the extensive study of thousands of mutants in bacteria and phages.

We have a few mutants available in amphibians and their number is increasing every year, but extremely few of them have been submitted to a biochemical analysis. This is unfortunate, since the already discussed *nucleolusless* (nu-o) mutant of *Xenopus* has played an exceedingly useful, if not decisive, role in the understanding of the intervention of the nucleolar organizers in rRNA synthesis and ribosome formation.

The only mutants that have been, to a very limited extent, analyzed from the biochemical viewpoint are the *o* and *cl* mutants of the axolotl. These two mutations lead to early lethality when they are in the homozygous state. *o* denotes *ova deficient* and *cl* denotes *cleavage arrest*. In the heterozygous condition, the *o* mutants reach the adult stage and are fertile, but many of the individuals die before they become adult. The homozygous *o/o* mutants die before or during gastrulation. Interestingly, they can be saved, as Briggs discovered, if one injects into the just fertilized egg the content of a germinal vesicle taken from a normal oocyte. The injection of this nuclear sap into one of the two first blastomeres enables the injected blastomere to develop further than gastrulation, while the other is blocked. Thus, the nuclear sap of the normal oocytes contains a factor, presumably a protein, that cannot be synthesized by the *o mutant* and that is necessary for development beyond the gastrula stage. Further studies on the chemical nature of this "corrective" factor will be awaited with great interest. We only know that it is present in the oocyte nuclear sap of a variety of amphibians and, thus, has no species specificity. The corrective factor is absent in the cytoplasm of ovarian oocytes, but is present in the germinal vesicle at all stages of *oogenesis*. RNA synthesis is very low in *o* mutants and it is, therefore, possible that the missing gene product is a karyophilic protein involved in RNA synthesis. Nuclear transplantation experiments have shown that nuclei taken from *o/o*-morulae support full development after transfer into enucleated, unfertilized eggs; but nuclei taken from *o/o* blastulae no longer support it. Thus, irreversible alterations occur in the nuclei between the morula and the blastula stages in *o/o* mutants.

The *nc* (no cleavage) mutant of the axolotl is also of interest for both molecular cytologists and embryologists: the eggs of this mutant can be fertilized, but they do not cleave because they are unable to assemble their large store of unpolymerized tubulin into spindle and aster microtubules. The mutation can be corrected, and repeated cleavages will occur after injection of centrioles (basal bodies of flagella) into the cytoplasm of fertilized *nc* eggs. What seems to be missing in these eggs is the capacity to assemble microtubule organizing centers (MTOC).

Comparable results have been found in *Drosophila* see also Chap. VII), in which a mutation called *deep orange* (*dor*) causes early death in the homozygous (*dor/dor*) embryos; the defect can be corrected by injecting cytoplasm from unfertilized normal eggs into *dor/dor* blastulae. It is not yet known whether, as in the *o* mutant of the axolotl, the repair factor is accumulated in the germinal vesicle during oogenesis. Other developmental mutations of *Drosophila* have been corrected by injection of cytoplasm from normal eggs.

It has been known for many years that the complete absence of the X chromosomes (*nullo-x* condition) results in deficiencies in early *Drosophila* development: gastrulation soon becomes abnormal and then stops. Small deletions in the X chromosomes may also lead to early arrest of development. It is now well established that the blocks in early development mentioned in this section are due to the fact that mutations, such as *o, cl, dor, nullo-x,* etc. affect the organization of the *cytoplasm* during *oogenesis:* the normal alleles of these genes control the organization (in gradients, for instance) and the chemical composition of the egg cytoplasm. Unfortunately, the products of these very important genes have not yet been identified.

Mention should be made of a few mutations affecting mouse development. The *Os* (oligosyndactylia) mutation, located on chromosome 8, causes digit fusion in the feet of heterozygous mice. In the homozygous state, it causes lethality between the seventh division and implantation, as a result of a mitotic block in metaphase. However, unlike the *nc* mutation in the axolotl, the *Os* mutation does not impair tubulin synthesis or microtubule assembly. It seems to be associated with a defect in spindle formation after the sixth division. Another dominant mutation, located on chromosome 2 [A^y (yellow)] is responsible for the yellow coat of the mutant mice; it also causes early lethality in the homozygous state. Cleavage is arrested before hatching from the zona pellucida and differentiation of the trophectoderm into trophoblast giant cells is totally impaired. Nothing is known about the biochemical target of the mutation.

The *t-complex,* located on chromosome 17, where many different mutant alleles are known, includes the H-2 locus which controls, in the adult mouse, the synthesis of the histocompatibility antigens responsible for graft rejection. Of particular interest, among the many *t*-mutants, are the t^{12}/t^{12} mutants where development stops already at the morula stage. Neither transcription nor translation are affected by this mutation until the embryo dies, but energy production must be somehow affected since neutral lipids accumulate in nuclear and cytoplasmic droplets.

5 "Transgenic" Mice and Teratocarcinoma

An old dream for many biologists has been to create new strains of animals with predetermined genetic changes: in other words, to introduce a selected gene into an egg and to witness its expression not only in the adult, but even in its offspring.

As we have seen, pure genes are now available in relatively large amounts due to progress in molecular biology. Parallel progress in embryology allows skilled experimenters to inject substances into one of the pronuclei of a fertilized mouse egg, to culture in vitro the injected eggs and to implant them into pseudopregnant females. If all goes well, fetuses will develop in their foster mothers and grow into adults, which will give birth to offspring after mating.

Injection of various genes into mouse pronuclei has been successfully achieved in recent years: if the gene coding for rabbit β-globin, for instance, has been injected into one of the pronuclei of a fertilized mouse egg, adult mice, in which all the cells contain several copies of the foreign (rabbit) gene, can be obtained. Still more important, these mice often transmit the gene to their offspring (or some of their offspring). These so-called transgenic mice fulfill the old dream of directing mutations, of manipulating Heredity at will. Success requires the integration of the foreign gene in the germ line cells of the recipient eggs and embryos.

Brinster has recently succeeded in performing similar experiments with "fusion-genes". A fusion-gene has been made of the promotor/regulator of the metallothionein I gene from the mouse and the structural human growth hormone gene. In mammals, metallothionein is normally produced by the liver and growth hormone by the anterior pituitary. When this fusion-gene is microinjected, the liver of the transgenic mice produces large amounts of human growth hormone, whereas their anterior pituitary does not. As shown in Fig. 68, such mice become giant mice. The same experiments are now feasible with eggs of domestic animals and could possibly lead to the economically important production of giant cows and pigs. From a more fundamental viewpoint, the most important feature of these experiments is that they offer a new way of studying gene regulation during mammalian development. It is now possible to learn

Fig. 68. Two 24-week-old sibling female mice. The mouse on the *right* contains an integrated new gene. The DNA fragment that was microinjected into mouse eggs is schematically represented as a *circle;* the *stippled* regions are exons and the *grey open boxes* are the introns of the human growth hormone gene. *MT* metallothionein; *pBR* portion of the vector (plasmid pBR 322). (Courtesy of Brinster et al. 1983; reprinted from *Science,* 1983, 222: 809–814 by copyright permission of AAAS)

much about the *tissue-specific regulation* of the expression of foreign sequences coding for globin, growth hormone, insulin, transferrin, metallothionein, immunoglobins, etc.

It has been shown that insertion of foreign genes in the germ line cells can induce mutations affecting development. The first case of *insertional mutagenesis* in mice was reported by Jaenisch: insertion of viral DNA sequences in mice resulted in a recessive lethal mutation, which causes, in the homozygous state, the death of the fetus at day 14 of pregnancy. Using the viral DNA as a probe, it was possible to show that the insertion lies in the 5′ end of the α1 (I) collagen gene; death of the embryos could be due to their incapacity to synthesize collagen. Insertional mutations have also been reported in mice carrying human growth hormone gene sequences; we have seen that insertion of a mobile genetic element into a *Drosophila* gene induces its mutation.

Similar experiments with human eggs have not yet been reported and it is obvious that very important ethical problems would arise if they were attempted. We all know that many "babies in test tubes" were born and that they grow normally. The methodology of artificial insemination, in vitro culture and reimplantation in a foster mother works as well for the human as for the mouse egg. "Transgenic men" are thus a theoretical possibility. Let us hope that if human eggs are ever manipulated (attempts to introduce new genes or to replace a "bad" gene by a "good" one), it will always be for the good and never for the evil of mankind. In this field, the moral responsibility of the scientist toward mankind is very heavy, not to say frightening.

Exciting material for both embryologists and oncologists is the *mouse teratocarcinoma,* a malignant tumor which originates spontaneously from the ovary or the testis, or can be induced by grafting 7-day-old embryos into the testis of male mice. Some teratocarcinoma cells differentiate into a monstrous embryo; others, called embryonal carcinoma cells, remain undifferentiated and are responsible for the malignancy of the tumor. If these "stem cells" are injected into a mouse, the recipient animal will be killed by a fast-growing tumor made of a variety of tissues as well as undifferentiated embryonal carcinoma cells.

The *pluripotentiality* of the embryonal carcinoma cells makes them attractive to embryologists because they share this property with the ectodermal cells of the inner cell mass. Particularly interesting is the fact that if a small number (three to five) of these cells are implanted into a normal mouse blastocyst, they are cured from their malignancy and can participate in normal development. Recently, teratocarcinoma cells have been used as vectors for *gene transfer* in mice: human β-globin genes have been introduced into mouse embryonal carcinoma cells by growing them in the presence of cloned human globin DNA sequences. The transformed cells may be introduced into normal blastocysts which are then grown in the usual way. The expected result is the production, in the second generation, of transgenic mice, thus of new strains with predetermined genetic changes (production of human β-globin) by another method rather than direct injection of a gene into a pronucleus.

As one can see, embryology is entering into a new era: fascinating, exciting, perhaps even dangerous discoveries lie ahead of us.

6 Conclusions

It is now time to draw a few conclusions before we leave the early stages of development and move into the much more complex area of embryonic differentiation. The work that has been summarized in the preceding chapters clearly shows that the chemical machinery for the initial steps of development has already been made during oogenesis. Not only ribosomes have been produced in such a high number as to allow development until a late larval stage (feeding tadpole) without further synthesis of rRNA or ribosomal proteins, but a great variety of mRNAs, containing a large array of information, have been stored in unfertilized eggs. New mRNA species are synthesized during cleavage, but they are not actually needed during this period which can normally proceed in the complete absence of RNA synthesis (except in mammals). The result of this economic, almost avaricious policy is that during the early stages of development, the importance of the cytoplasm largely overwhelms that of the nucleus. The critical point is the end of cleavage; afterward, gastrulation movements, and neural induction become impossible without fresh information, resulting from gene activation and transcription.

But one should not forget that the fundamental, basic processes that control final development occur during these very early stages. Polarity, which will control cephalocaudal differentiation, formation of the grey crescent in amphibian eggs, (an immediate consequence of fertilization, responsible for the dorsoventral organization of the egg), germinal localizations in the "mosaic" eggs and embryonic regulation, all occur at stages in which the genes are almost entirely repressed and mRNA synthesis is hardly detectable.

If one returns to the different theories of embryonic differentiation which have been presented in Chap. III, one cannot escape the conclusion that the ideas expressed by Morgan, almost 50 years ago, remain valid as far as the early stages of development are concerned. His concept that gene activity is controlled, at these early stages, by the properties of the heterogeneous cytoplasm which surrounds equipotential cleavage nuclei has been substantiated by many experiments (nuclear transplantations, in particular). It is now clear that the nuclei of adult, differentiated cells and those of zygotes possess the same genes, but cytoplasmic factors present in oocytes and eggs affect their expression. The spectrum of proteins synthesized under the control of an adult nucleus is not the same whether it has been injected into an oocyte or it lies in its own cytoplasm. The identity of the *cytoplasmic factors* present in oocytes and eggs, which allow the reprogramming of injected nuclei, remains unknown. If one may venture a guess, these factors might well by karyophilic proteins, which have the remarkable capacity to move quickly from the cytoplasm into the nuclei where they soon accumulate and may bind to DNA. Detailed analysis of the proteins present in the oocyte nucleus and in cleaving egg nuclei is an important and urgent task for the molecular embryologists of tomorrow.

The cytoplasmic factors that have just been discussed certainly continue to play their role in later stages of development, when the cells undergo their typical differentiation from the morphological, physiological and molecular viewpoints. But their importance becomes less and less evident as development proceeds. The sequential, selective activation of individual genes or groups of genes now appears in the forefront, as we shall see in Chap. XI.

CHAPTER XI
How Cells Differentiate

1 General Introduction

The basic problems involved in embryonic differentiation have already been discussed at the beginning of this book. A typical question, which we have already asked before, is the following: How is it possible that red blood cells, which have synthesized large amounts of a very specific protein, hemoglobin, appear only in a very localized area of the embryo, at a very precise stage of its development? Figure 69 gives a good example of embryonic cell differentiation; it is a section through a young tadpole and shows, side by side, two types of highly specialized cells, which have just undergone embryonic differentiation. *Cartilage* cells and *muscle* cells lie side by side and are easily recognizable by their morphology. The cartilage cells are surrounded by an amorphous matrix made of special glycoproteins, called *chondroproteins;* the muscle cells contain fibrils (the myofibrils), which are made of entirely different proteins, in particular, the contractile proteins *myosin* and *actin.* The structure, the function and the biochemical composition of the cartilage and muscle cells are thus very different and specific for each type of cell. Yet, if the section had been made a couple of days earlier, it would not have shown such highly differentiated cells; histological studies of the embryonic development, during this short 2-day period, would have shown that cartilage and muscle cells have *the same* mesodermal origin. In a group of mesodermal cells, which all look alike, if not identical, some cells will form glycoproteins in large amounts and become cartilage; others will synthesize contractile proteins and differentiate into muscle cells.

Since we know, due to the nuclear transplantation experiments of Gurdon, that all the nuclei of the tadpole and even the adult contain the same genes as the fertilized egg (the zygote), differentiation can only be explained by assuming that in muscle cells, certain genes that direct the synthesis of the contractile proteins are active, while those that direct the hemoglobin and chondroproteins are silent. In the same way, the hemoglobin genes must be active in the differentiating red blood cells and the chondroprotein genes must be functioning in the young cartilage cells. Thus, the genes involved in the synthesis of large amounts of the other specific proteins (which Holtzer called "luxury" proteins) must be inactive in these cells. On the other hand, the genes that control the synthesis of proteins which are needed by *all* cells must necessarily always be active. Any cell, whether it is a muscle cell, a red blood cell, a cartilage cell, etc.; requires energy and must

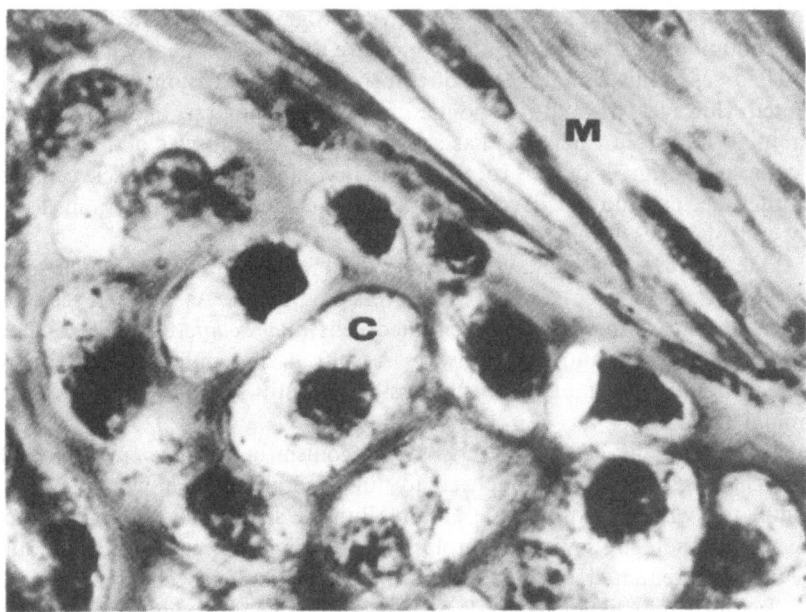

Fig. 69. Cell differentiation in an amphibian embyro. *C* cartilage cells; *M* muscle cells

possess all the enzymes needed for oxidative phosphorylations (cytochrome oxidase, dehydrogenases, etc.). Differentiated cells contain variable amounts of these enzymes; Holtzer called them the "housekeeping" proteins, and they must be present if the cell is going to survive. Thus, the problem of cell differentiation is largely a problem of *gene regulation.* This important question has been discussed in Chap. III, where we have seen that many different mechanisms may operate in the control of gene activity. During cell differentiation, where one or a few genes coding for specific luxury proteins are selectively expressed, there is little doubt that the main control takes place at the *transcriptional* level. As we have seen, "active chromatin" displays several peculiarities: high sensitivity to DNase digestion, frequent undermethylation of the DNA molecule and quantitative or qualitative changes in the DNA-binding proteins are well documented to day. But this is still a crude approach and we do not yet know how specific genes are turned on or off in a very precise and delicate way. We do not even know by which mechanisms the hemoglobin genes are undermethylated in the chromatin of red blood cell precursors and fully methylated elsewhere. How apparently undifferentiated cells become "committed" to differentiate along a given developmental pathway (erythropoiesis, myogenesis) remains, despite many efforts, an unsolved problem.

One of the hypotheses proposed to explain cell differentiation is that of "quantal mitoses", cell divisions which would give rise to two *non-identical* cells: one of them would remain an undifferentiated *stem cell,* the other would be committed to differentiate according to a given program. By repeated cell divisions, this committed cell would ultimately give rise to a *clone* of nearly identical differentiated cells. This clonal theory of embryonic differentiation is now widely accepted. However, it should be pointed out that the analysis of allophenic (tetraparental) mice (see Chap. IX) has shown that all the muscles of a mouse derive from a small number of cells committed to become myocytes (muscle cells) in the somites, i.e. they do not derive from a single clone of cells, but from a few independent clones of cells.

One usually speaks today of *terminal* differentiation, a term which seems to imply that cell differentiation is an irreversible process. As we shall see, there is at least one case in which differentiated cells may *dedifferentiate* and then redifferentiate along another pathway. A shift in the differentiation program, called *transdifferentiation,* is thus a possibility, but it remains a rare event in the animal kingdom.

It is often believed that there is an antagonism between *cell division* and *cell differentiation;* indeed, mitotic activity, in general, slows down or even stops completely when cells fully differentiate. But one should not be too dogmatic on this point. There are several examples in which dividing cells have been found to be fully differentiated, as evidenced by the production of specific luxury proteins.

Two different biological systems can be used for the study of cell differentiation: developing embryos and in vitro cultures of embryonic cells committed to differentiate along a particular developmental pathway. The two systems are adequate for the analysis of cell differentiation, but both have their advantages and disadvantages. Whole embryos represent a more "natural" system than cultured cells; although the latter are easier to work with, the results obtained with them do not necessarily apply to whole embryos. This is not surprising since, in embryos, in contrast to cultured cells, cellular movements, cell-to-cell interactions, morphogenetic gradients and inductions play a decisive role. When embryos are dissociated into individual cells, these cells lose the precise localization they had in the morphogenetic gradient where they interacted with their neighbors. The result is the loss of what has been called *positional information.* This, as well as the poor permeability to drugs of whole embryos from certain species (amphibians, in particular), explain why some agents which, as we shall see, suppress cell differentiation in cultured embryonic cells have very little effect on intact embryos.

It is impossible to give a full account, in a few pages, of all the work that has been done and is being done on cell differentiation. We shall concentrate on *embryonic* differentiation and speak more superficially of the studies made by the workers who use tissue cultures as their main method.

Before we come to the presentation and discussion of the known facts, a last word of introduction should be said. The study of cell differentiation has more than an academic interest. It is a problem which is of fundamental importance for mankind, because cancer is, to a large extent, a disease in which the cells divide, but fail to differentiate. The solution of the cell differentiation problem might lead sooner or later to a solution of the cancer problem. If malignant cells could be made to differentiate, they would stop dividing and would perhaps no longer be malignant.

2 Specific Properties of Cell Membranes

Already at the gastrula stage, as shown by Holtfreter, cells of the different parts of the embryo differ in their *tissue affinities*. As shown in Fig. 70, if cells of the ectoblast and the entoblast are joined together in the same explant, they separate from each other after a few hours of culture. They have *negative affinities*, since they reject each other. On the other hand, the mesoblast cells stick to either ectoblast or entoblast when they are placed in contact; they have *positive affinities* for the cells of the two other cell layers. Interestingly, treatments with actinomycin or cycloheximide modify the ectoblast tissue affinities: their acquisition or mainte-nance thus requires the synthesis of macromolecules.

Comparable findings have been made with cells isolated from much older embryos (Fig. 71). For instance, Moscona dissociated cells from the retina of chick and mouse embryos and mixed them with other dissociated cells, coming from the kidney of the same embryos. He found that retina cells recognize each other and form a *chimaeric* retina, made of mouse and chicken cells; in the same way, the dissociated kidney cells form *chimaeric* kidney tubules, made of cells originating from the two species of embryos. In this case, the cells recognize the other cells which belong to the same tissue and are members of the same family.

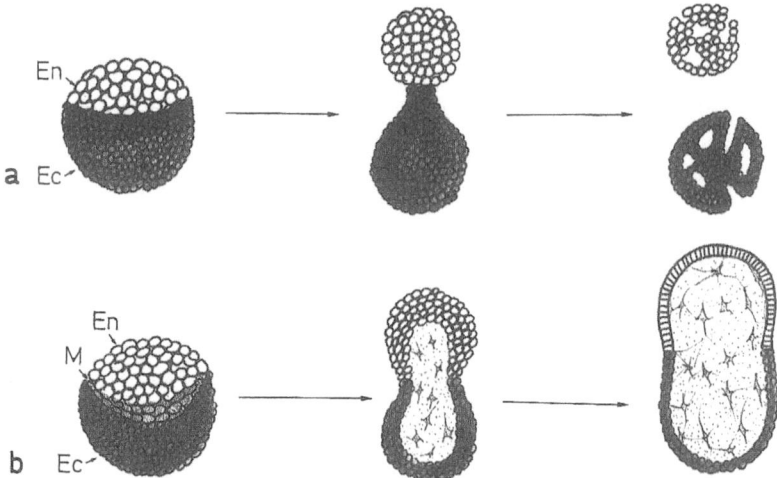

Fig. 70 a, b. Tissue affinities. **a** Ectoderm (*Ec*) and endoderm (*En*) show "negative" affinity; they quickly separate from each other and form unorganized gut and epidermic cells. **b** If a piece of mesoderm (*M*) is inserted between the two fragments, ectoderm and endoderm no longer separate; they both have a "positive" affinity for mesoderm; a vesicle containing mesenchymatous cells is obtained (J. Holtfreter)

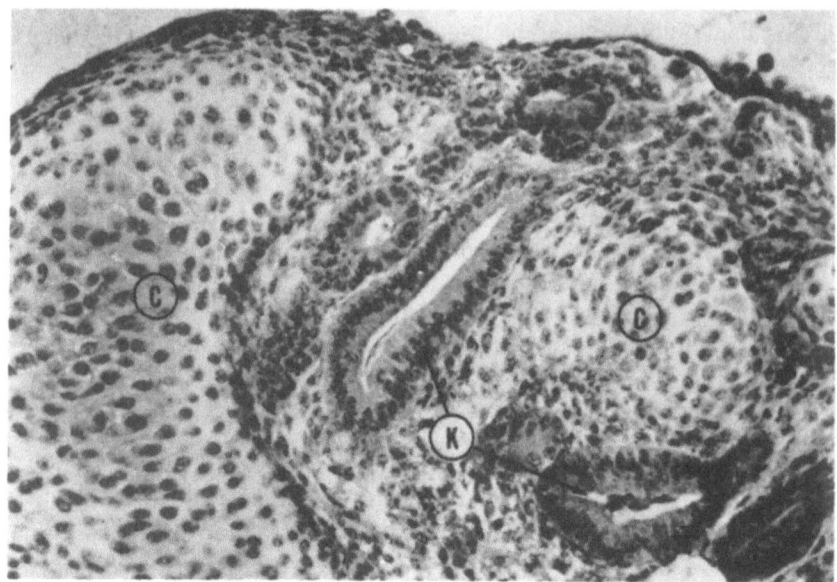

Fig. 71. Aggregate of mesonephric and cartilage cells of a chick embryo. After 7 days of culture, the two types of cells have segregated and reconstructed nodules of cartilage (*C*) and kidney tubules (*K*) (Monroy and Moscona 1979)

Cell surface

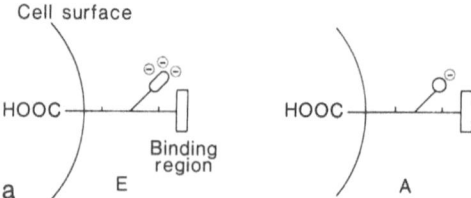

Fig. 72 a, b. Schematic model for N-CAM (neurone- specific cell adhesion molecules). The sialic acid-free binding region is represented by the *rectangle* and the sialic acid-rich sugar moiety by an *oval;* the possibility of more than one site of sugar attachment is shown by *vertical lines*. **a** The embryonic (*E*) to adult (*A*) conversion removes large amounts of sialic acid. **b** Adhesion of two homologous cells (neurones) by association of ligand molecules (Edelman 1983; reprinted from *Science*, 1983, 219: 450–457 by copyright permission of AAAS)

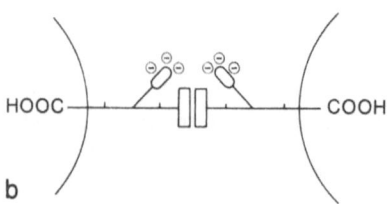

Many attempts are being made today to identify the molecules responsible for specific cell-to-cell adherence in embryonic tissues and in lower invertebrates, such as the sponges (where reaggregation of dissociated cells is easy to obtain). These molecules (called *CAM* for cell adherence molecules or Cognins) display tissue specificity: L-CAM from liver favors the reaggregation of liver, but not of nerve cells. In contract, N-CAM from the nerve cells acts specifically on the re-aggregation of neurones and has no effects on liver cells. All these molecules are large acidic glycoproteins, with a molecular weight of more than 100,000. Figure 72 shows schematically the relationships between N-CAM, a nerve cell and the outer medium. N-CAM is formed of two closely related polypeptide chains with molecular weights of respectively, 140,000 and 170,000. At the end of the amino terminal region is a binding site for another N-CAM molecule, while the other end (carboxyl terminal region) penetrates into the cell through the cell membrane. Specific cell adhesion molecules are already present at early stages of development and are responsible for the reaggregation of dissociated sea urchin blastula cells and for compaction of 8-cell-stage mouse embryos as seen in Chap. IX. It has been recently proposed by Edelman that CAM genes control the morphogenetic movements required for induction.

An additional word should be said about the immunity reactions which follow the grafts made in amphibian embryos. These grafts are very well tolerated when a piece of tissue (an organizer, for instance) is grafted in a host of another species. A normal head, with eyes, nose, etc., can be induced in the host, but sooner or later, this secondary head is rejected and destroyed. Incidentally, such experiments show that the inducing capacity of an organizer does not display any species specificity, but the response of the host to the inductor is species-specific and thus gene-determined. A frog organizer grafted into a newt embryo induces newt nervous tissue.

From the viewpoint of the immuno-response, there is a difference between amphibian and mammalian embryos. If a foreign tissue is grafted in the latter, it is not rejected even when the immuno-system has been built up after birth, (acquired tolerance).

3 Effects of Tissue Extracts on Cell Differentiation

Many efforts have been made in the past with the hope of isolating substances from differentiated tissues, which would induce the tissue-specific differentiation of undifferentiated cells. These efforts have so far been frustrating and the limited effects obtained should be ascribed to the presence of many factors in tissue extracts, which stimulate or retard cell proliferation and, as a result, may affect cell differentiation in an indirect way. Substances present in adult tissues which inhibit mitotic activity of homologous embryonic cells in a tissue-specific way are called *chalones*. For instance, an extract of adult kidney reduces the number of mitoses to a greater extent in embryonic kidney (pronephros) than in all other tissues of an amphibian larva. Many different chalones have been described;

they seem to be very heterogeneous molecules, since their molecular weights may vary from less than 10,000 to more than 100,000.

It is beyond the scope of this chapter to discuss the numerous substances (*growth factors*) which stimulate cell division and may indirectly slow down differentiation. Many of them bear some similarities with the pancreatic hormone *insulin* and act by binding to specific receptors present on the cell membrane. The growth factor-receptor complex is internalized by endocytosis and partly broken down; the growth factors ultimately stimulate DNA replication in the cell nucleus. The best known of these growth factors is the *ectodermic growth factor* (EGF); this small polypeptide (6000 daltons, 53 amino acids) exerts many early effects on fibroblasts (increases in Na^+ uptake, intracellular pH, amino acid and glucose transport, etc.) which finally lead to the stimulation of nuclear DNA synthesis.

4 Immunological Studies on Embryonic Differentiation

Immunological (serological) methods are exceedingly valuable for the detection of *specific proteins*. If a protein (an *antigen*) is injected into an animal, a rabbit for instance, this animal will produce in its serum an *antibody*. This is a protein which is called an *immunoglobulin* and which reacts in a highly specific way with the antigen. The antigen-antibody complex is less soluble than the antigen and the antibody. The presence of the antigen can thus be detected by its immunological precipitation after addition of the antibody. If one binds to the antibody, by chemical means, a fluorescent dye or a radioactive marker, it becomes possible to detect, under the microscope, the cells which contain the antigen. The present technology of *monoclonal antibodies* production makes antibodies of exquisite selectivity available; they are produced by *hybridomas,* which result from the fusion of a malignant myeloma cell and a cell taken from the spleen of an animal which has been immunized against the substance of interest.

Early work in which embryos crushed at various stages of their development were used as antigens showed that new antigens appear at amphibian gastrulation and that others can be detected at later stages of development. This merely confirms the data that physical methods for the separation of newly synthesized proteins had shown. It is probable that the use of monoclonal antibodies would lead further, but the most important application so far of immunological methods to embryology has been the analysis of the formation of the eye *lens* proteins (the α, β and γ-crystallins).

The *lens* is induced by the optic cup, at the expense of the ectoderm, at a late neurula stage. Immunological analysis shows that the lens antigens can be detected in the head of the young neurula. They are present in all cells, but in low concentrations. When the lens is induced, the specific lens antigens become detectable in the still undifferentiated lens in higher concentrations than elsewhere. Their concentration considerably increases at the time of cell differentiation, characterized by the appearance of fibers, made of crystallins, in the center of the

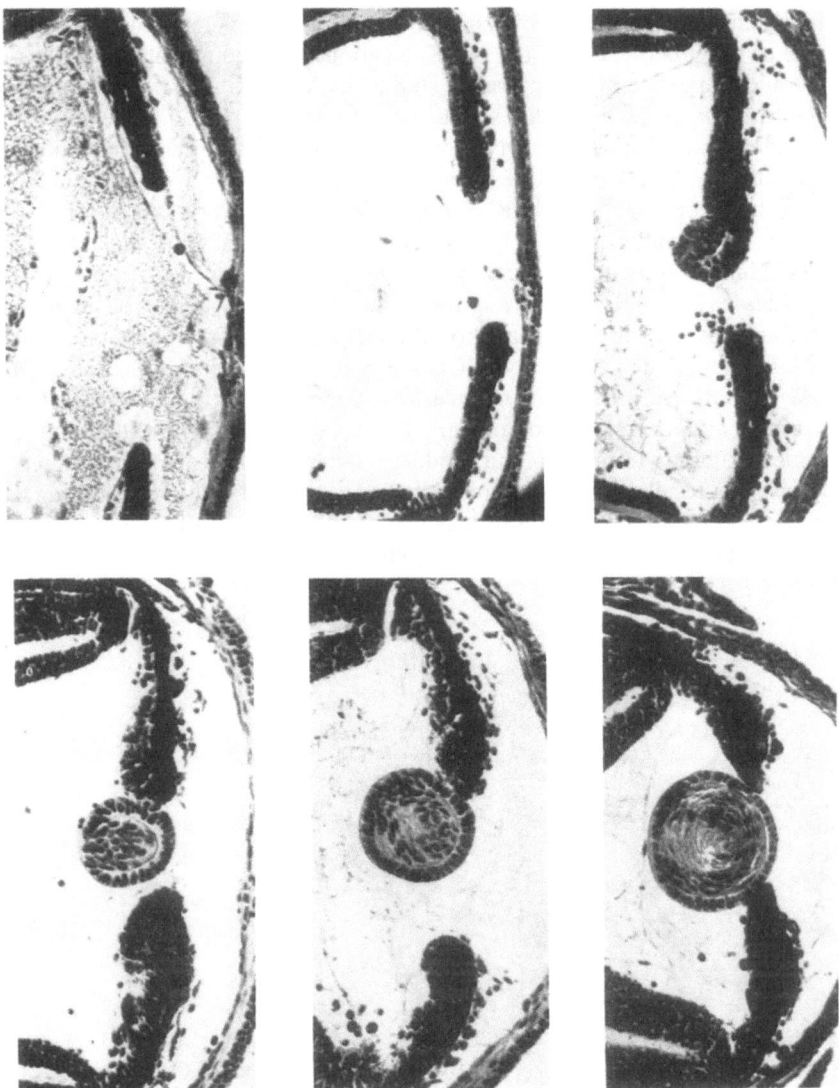

Fig. 73. Stages of Wolffian regeneration of the lens from the superior rim of the iris (Sato 1930)

embryonic lens. This is due to a selective activation of the crystallin genes under the influence of the inducing retina cells. There is thus a progressive restriction in the synthesis of the lens protein from the whole head to the eye lens only during development.

Another system of interest is the so-called *Wolffian regeneration of the lens* in adult amphibians or advanced tadpoles. As shown in Fig. 73 when the fully differentiated lens is removed by surgical operation, a new lens forms at the expense of the superior rim of the iris. The various biochemical phases of lens regeneration have been very accurately studied by Yamada. The first event is a *dedifferentiation* of the iris cells, which lose their black pigment (melanin). Depigmentation is followed by DNA synthesis and cell multiplication. When mitotic activity stops, a period of intensive RNA synthesis sets in. Large nucleoli make their appearance and rRNA synthesis becomes very conspicuous. This phase of RNA synthesis is very important for the success of the regeneration. The latter is inhibited by actinomycin, which blocks RNA synthesis. The mRNAs for the *crystallins* are also synthesized during this period and these proteins become detectable, by immunological methods, immediately after the phase of extensive RNA synthesis. The various crystallins are synthesized one after the other. They are found, at first, in both the cytoplasm and the nucleus; the latter degenerate afterward and completely disappear. These experiments show that inducing agents originating from the retina produce a derepression of the genes present in the iris. As a result, synthesis of the crystallins is induced in the regenerating lens. Cell differentiation (formation of lens fibers) and synthesis of the crystallins are closely linked and go hand in hand. Wolffian regeneration of the lens is an in-

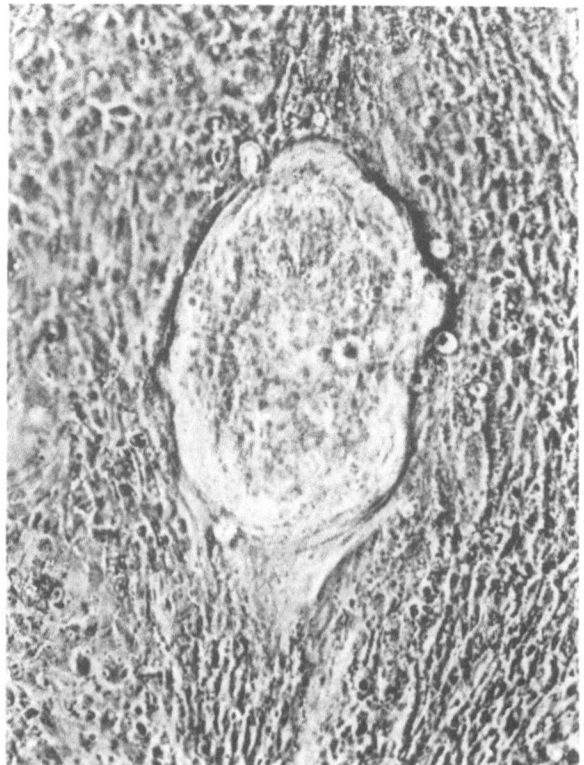

Fig. 74. A lentoid body appeared in a culture of a cell deriving from retina (Eguchi and Okada 1973)

teresting case of *transdetermination*, the phenomenon first discovered by Hadorn in transplanted imaginal disks of *Drosophila* larvae. A similar phenomenon, which has been called *transdifferentiation*, can be obtained with cultured neural retina embryonic cells: some of the pigmented cells lose their pigment and differentiate into lens cells. Small lenses, called lentoids (Fig. 74) make their appearance in the cultures. As in Wolffian regeneration, inactive crystallin genes are activated: crystallin mRNAs are produced in large amounts and the corresponding proteins accumulate in the lentoids.

5 Enzyme Synthesis and Embryonic Differentiation

It has often been suggested that differentiation might be due to the induction of enzyme synthesis resulting from the appearance, in a region of the embryo, of the substrate for the enzymatic reaction, or to the repression of enzyme synthesis by the localized accumulation of the end product of the enzyme reaction. Such control mechanisms of enzyme synthesis operate efficiently in bacteria and the elucidation of their genetic and molecular bases is at the root of the famous Jacob-Monod model of gene regulation in bacteria. But eggs are not bacteria, and all efforts made to demonstrate substrate-induced synthesis in embryos have been unsuccessful.

The presence of many enzymes in unfertilized eggs makes an enzymological approach to the study of cell differentiation difficult: some of the enzymes synthesized during oogenesis are kept ready for use at much later stages of development. However, this approach has been valuable in a few cases, as we shall now see.

Several *hydrolytic enzymes* are known to increase in activity in the digestive tract when it becomes functional. This is true for amylase, proteases, phosphatases, etc. in sea urchin plutei as well as in amphibian tadpoles and chicken embryos. In such cases, enzyme synthesis is clearly linked to the acquisition of a physiological function.

The activity of *cholinesterase*, an enzyme which plays an essential role in the nervous system, is already important in the unfertilized amphibian eggs. It increases considerably when, after neural induction, the nerve cells begin to differentiate into typical neurones which are physiologically active. Similar changes in cholinesterase activity have been recorded in the cells of a tumor of the nervous system, *neuroblastoma*. The tumor cells have a low cholinesterase activity, but, if they change into more typical neurones (provided with nerve fiber expansions) as the result of a change in the in vitro culture conditions, cholinesterase activity very markedly increases as the result of the synthesis of new enzyme molecules.

Changes in enzyme synthesis, correlated with modifications of the outer medium, have been recorded in other systems as well. For instance, *glutamine synthetase*, an enzyme which is also involved in nervous system metabolism, is synthesized, at a given time of development, by the retina of the chick embryo. If the retina cells are dissociated, and then allowed to reaggregate, glutamine synthetase synthesis is accelerated by several hours. Still earlier synthesis of glutamine synthetase is obtained when the retina cells are treated by cortico-steroid hormones. The use of macromolecule synthesis inhibitors suggests that glutamine synthetase synthesis is regulated at both the transcriptional and trans-lational levels.

It is, of course, impossible to give here a complete summary of all the work that has been done on enzyme synthesis during development and we shall close the discussion with a last example, that is, the synthesis of the *pancreatic enzymes* (trypsin, chymotrypsin, amylase, etc.) which has been the subject of very interest-ing studies by Rutter. The pancreatic "acini," forming the glandular site of diges-tive enzyme synthesis, are induced by the surrounding *mesenchyme* (the still un-differentiated mesodermal tissue). Their induction is of the "transfilter" type: it is successful even if the reacting and the inducing tissues are separated by a milli-pore membrane. Rutter succeeded in isolating a protein from the filters, which certainly plays a great role in the induction. This protein induces in the reacting cells the synthesis of a particular kind of low molecular weight DNA. This syn-thesis is followed by the replication of the main DNA, which leads to sustained mitotic activity. Only when a sufficient number of cells becomes available does RNA and then protein synthesis occur. The various pancreatic enzymes are syn-thesized in a constant order, each following the other. There are no indications that enzymes which serve in an identical metabolic pathway, protein or sugar metabolism, for instance, are synthesized together. This speaks against the exis-tence of an "operon," similar to those of the bacteria. It is clear that the structural genes coding for the different pancreatic enzymes are activated in a sequential and individual manner, since there is a concomitant increase in their respective mRNAs: for instance, the content of the pancreatic cells in amylase mRNA in-creases 400 times during their differentiation. In this case, the main control of en-zyme production is clearly at the transcriptional level.

Transcriptional controls are particularly clear when organs respond to *hor-monal* stimulation by the production of large amounts of a specific protein. For instance, steroid hormones induce the secretion of very large quantities of oval-bumin in the chick oviduct, and of vitellogenin in the amphibian liver. The hor-mone binds to a receptor in the target cell; in the nucleus, the hormone-receptor complex activates selectively the ovalbumin or the vitellogenin genes, probably by binding to them. As a result, chromatin in the structural gene and its flanking regions takes the loose conformation of "active" chromatin. The gene is now ex-posed to RNA polymerase activity and the final result is an enormous production of the corresponding mRNA, which is immediately translated into protein; the latter is finally excreted. While there is less than one molecule of ovalbumin mRNA in each cell of the unstimulated chick oviduct, thousands of molecules per cell are found after steroid hormone administration. As we have seen, the re-sponse of the amphibian oocyte to progesterone is fundamentally different since gene transcription has very little, if any, importance for protein synthesis. This process is controlled at the translational level.

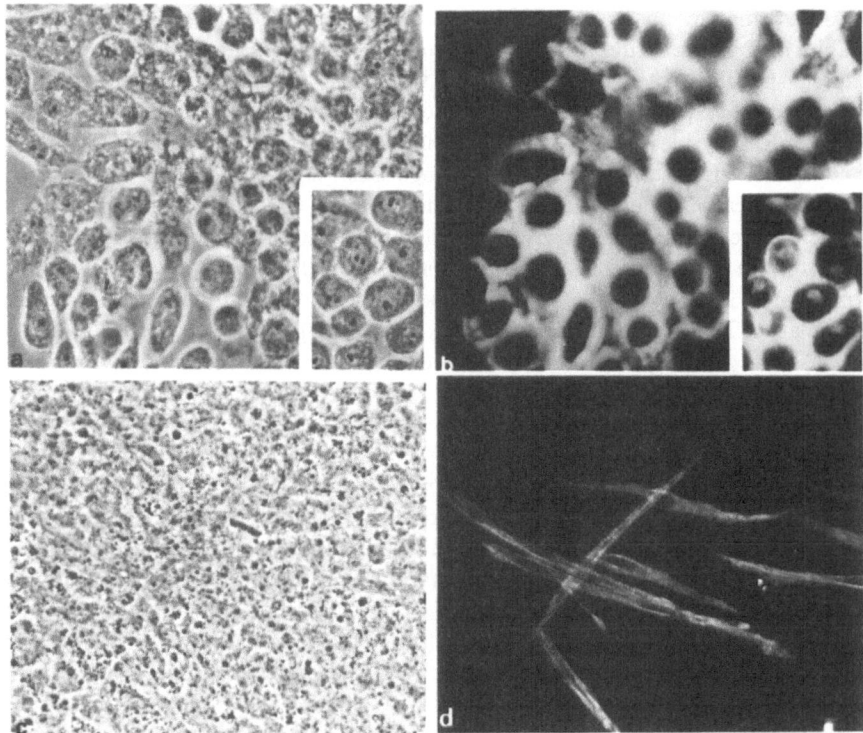

Fig. 75a–d. Cell differentiation in vitro: chondrogenesis (**a, b**) and myogenesis (**c, d**). Phase contrast (*left*) and immunofluorescent (*right*) micrographs of monolayer cultures of chick embryo chondroblasts (**a, b**) and myoblasts (**c, d**). Differentiating chondroblasts were stained with an antiserum raised against the chondroitin sulfate-rich cartilage proteoglycan (**b**). *Inset:* bright fluorescence on the Golgi complex. Differentiating myoblasts were stained with antilight meromyosin antibody (**d**). (Courtesy of Prof. H. Holtzer)

6 Differentiation of Cultured Embryonic Cells

6.1 A Brief Description of a Few Biological Systems

Numerous studies on the in vitro differentiation of undifferentiated, but already "committed" embryonic "blast" cells have led to the conclusion that the key event is the selective activation and expression of a small number of specific genes.

Myogenesis (muscle differentiation) is one of the favorite systems for the in vitro study of cell differentiation. Undifferentiated *myoblasts* fuse together in culture and form *myotubes*, whereby the latter finally differentiate into contractile fibers, and simultaneously, several muscle-specific proteins (in particular, muscle actin and myosin) make their appearance (Fig. 75). DNA synthesis stops when myoblasts fuse together. Experiments with actinomycin have shown that myoblasts synthesize unstable mRNAs, while differentiated muscle cells contain stable mRNA molecules. Large amounts of the messengers coding for the muscle-specific proteins are produced when the myotubes differentiate into muscle cells. In contrast, the bulk of the mRNA population (about 17,000 different mRNAs) does not vary greatly during myogenesis.

Another interesting system is *chondrogenesis* (cartilage formation), in which embryonic *chondroblasts* differentiate into *chondrocytes* which incorporate inorganic sulfate and build up an extracellular matrix of sulfated proteoglycans. Chondroitin sulfate and chondroproteins are typical markers of cartilage differentiation (Fig. 75). Soluble factors released by in vitro growing chondrocytes stimulate differentiation of chondroblasts into cartilage.

There is growing interest in the differentiation of some fibroblast strains (preadipocytes) into *adipocytes* (fat cells). Treatment with insulin triggers or accelerates the accumulation of triglycerides in the cytoplasm which becomes filled with a large lipid droplet; simultaneously, the cytoskeleton breaks down.

Many papers have been devoted to *erythropoiesis,* the formation of red blood cells, which is characterized by the synthesis of an easily detectable protein, hemoglobin. Erythropoiesis is stimulated by the addition of *hematopoietin,* a protein hormone made mainly by the kidney. The molecular organization of the α- and β-globin genes coding for the two globin chains of hemoglobin is very well known: we can follow, at the gene level, the shift from embryonic to fetal and finally adult hemoglobin. We know that the α- and β-genes lie on different chromosomes, and that there is no amplification of these genes when large quantities of hemoglobin are synthesized during erythropoiesis. We also have a growing knowledge of the multiple molecular causes of the *thalassemias,* hereditary diseases in which there is an unbalance in the synthesis of the α- and β-globin chains. It is the dream of many molecular biologists to cure these diseases by transfer of the correct gene into the deficient genome of the patient (gene therapy); success is not very far ahead. Finally, a large amount of work is being done on Friend's murine *erythroleukemic* cells (MEL cells): these cells are malignant because they are infected by Friend's leukemia virus. They grow continuously in culture, but do not differentiate. However, differentiation, evidenced by the synthesis of hemoglobin and a few other marker molecules, can be induced by treating MEL cells with simple chemical agents (dimethylsulfoxide, butyrate, etc.). In all these cases, whether hemoglobin synthesis occurs spontaneously or is induced experimentally, there is a close parallelism between the number of globin mRNA molecules present in a cell and the amount of hemoglobin it synthesizes. Thus, during erythropoiesis, a small number of genes are activated sequentially.

Among the other systems in which cell differentiation can be followed in vitro, mention should be made of the formation of *melanocytes* (pigment cells) and *neurones* (nerve cells). Melanocytes contain a brown pigment, called melanin, which makes these cells easily recognizable under the microscope.

Neurones possess *neurites* (axones and dendrites), have typical electrophysiological responses when they are excited and contain a number of specific proteins. The outgrowth of neurites is stimulated by a *nerve growth factor* (NGF) which has been well characterized. Cell lines derived from brain tumors (neuro-blastomas, neurogliomas) grow actively in culture; these cells are undifferentiat-ed, but they can be induced to differentiate and to form neurites by treatments with a number of physical and chemical agents which tend to slow down their growth.

It has been possible to isolate, from teratocarcinomas, embryonal carcinoma cells (ECC) of different origins and to produce a large number of established cell lines. Some of them remain pluripotent after prolonged cultures, while others be-came restricted in their differentiation potentialities. Sometimes they gave rise to a single committed cell type, which is the case for the so-called neuro-terato-carcinoma stem cells which differentiate in vitro only in neurones. Another ECC line, F9, differentiates into visceral endoderm if treated with retinoic acid (a vita-min A derivative) and into parietal extraembryonic endoderm if treated with a retinoic acid and cAMP mixture. Differentiation of F9 cells into visceral en-doderm cells is accompanied by the activation of the α-fetoprotein gene. This gene remains silent in F9-derived parietal endoderm cells, which synthesize in-stead, plasminogen activator, cytokeratin filaments, laminin and type-IV pro-collagen.

6.2 Experimental Analysis of Cell Differentiation in Culture

6.2.1 Cell Fusion

A very interesting approach has been developed for the study of cell differen-tiation: it is the technique of *cell fusion,* also called *somatic hybridization* (Fig. 76), which was first devised by Ephrussi. The properties of the cell membrane can be altered by treatment with a hemolytic virus, called the Sendai virus, even when this virus has been inactivated and is unable to multiply in the infected cells, or with polyethylene glycol. These changes in membrane properties allow the easy fusion of two (or more) cells of the same or different types. As Harris has done, when a nucleate hen red blood cell is fused with a human cancer cell (HeLa cell), very remarkable changes occur. The nucleus of the red blood cell (erythrocyte) was, before fusion, in a completely inactive, repressed form. It synthesized neither RNA nor DNA; the HeLa cell nucleus, on the other hand, is extremely active in both DNA and RNA synthesis. When the two cells are fused together, a

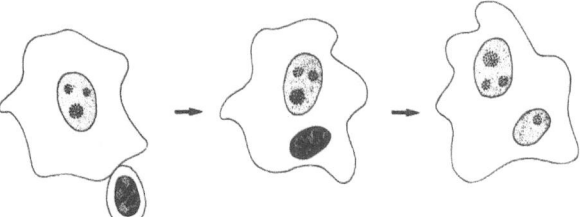

Fig. 76. Fusion of a chicken red blood cell (small, with a condensed nucleus) with a mammalian cell: the re-activated red blood cell nu-cleus forms a nucleolus

heterokaryon is produced. There are two nuclei, of different origins (hen and man) in a common cytoplasm. Analysis of this somatic hen-man hybrid shows that the erythrocyte nucleus is quickly *reactivated*. It becomes capable of synthesizing DNA and RNA. The hen erythrocyte nucleus contained no visible nucleoli before the cell fusion; after cell fusion, they become very conspicuous and RNA-rich. The nucleolar organizers are now active again. At that same time, it becomes possible to detect, by immunological methods, the formation of chick-specific antigens in the cell membrane of the somatic hybrid. The HeLa cell *cytoplasm* has thus exerted a positive, derepressing effect on the previously wholly inactive chick nucleus. This effect is entirely similar to the one we met before when the experiments of Gurdon on unfertilized *Xenopus* eggs were described. Injection of a nucleus taken from an adult organ (brain, liver), where there is no longer any DNA synthesis, into young cytoplasm is followed by a resumption of DNA synthesis. Clearly, the same kinds of positive control that the egg cytoplasm exerts on the injected adult nuclei still operate in active adult cells.

This elegant technique of cell fusion has been used by Ephrussi in the study of cell differentiation. He fused together *melanocytes* and still undifferentiated embryonic cells (*fibroblasts*) and cultured these hybrids. The result, in this case, was that the undifferentiated cell exerted a *negative* control on the differentiated one. The hybrid between the two cells soon lost not only melanin, but even the capacity to synthesize tyrosinase. Similar results were obtained with liver and nerve cells. The former are no longer capable of synthesizing a number of enzymes and blood proteins after fusion with an undifferentiated cell. Nerve cells no longer synthesize a specific protein (called S 100) of the nervous system after fusion with embryonic fibroblasts. The control exerted by the undifferentiated cell on differentiated functions is not *always* negative but this is an exception to the general rule. Studies on the fusion of malignant and normal cells have shown that the situation prevailing in somatic hybrids can be very complex. Malignancy usually behaves like a recessive character, but the interpretation of the results is complicated by the fact that cell fusion followed by culture is often followed by the elimination of part of the chromosomes (this also often happens, as we have seen, in the ordinary sexual hybrids). Experiments, in Ephrussi's laboratory, have shown that if the melanocytes contain twice as many chromosomes as the fibroblasts, the ability to produce melanin is no longer lost in the hybrid between the two cells. This clearly shows the importance of "gene dosage" for the expression of the phenotype (melanin production).

Recent studies indicate that the rule, after cell fusion and culture, is the following: the differentiated phenotype (formation of pigment, liver enzymes, S 100 protein, etc.) is lost soon after cell fusion; but it reappears when a certain number of chromosomes have been lost during cell divisions in the hybrid line. The differentiated phenotype is first extinguished, then re-expressed. In order to gain some insight in the relative roles of the nucleus and the cytoplasm in the phenotypic expression of differentiation, cytoplasm from enucleated fibroblasts (called *cytoplasts*) has been fused with intact cells. In such *"Cybrids"*, extinction of the differentiated character does not last more than 12 to 20 h; the differentiated phenotype is restored 48 h after fusion. This experiment shows that a cytoplasmic factor exerts a negative control on cell differentiation, but that this factor has a short life and that it is not renewed in the absence of the nucleus. Since enucleation induces the differentiation of neuroblastoma cells (neurite formation), it is likely that the nucleus contains or produces factors which inhibit cell differentiation. This is another case of a negative control exerted by the nucleus on the cytoplasm. It is likely that extinction of the differentiated phenotype is due to changes in the molecular organization of chromatin. But it is too early to draw strong and general conclusions in a fast moving and complex field. We still know very little about the molecular interactions which take place between nucleus and cytoplasm in somatic hybrids and cybrids.

6.2.2 Effects of Bromodeoxyuridine (BrdUr) on Cell Differentiation

There is another interesting approach to the analysis of cell differentiation: Holtzer and colleagues found that an analogue of thymidine, bromodeoxyuridine (BrdUr) specifically inhibits cell differentiation in most cultured cells. At low concentrations, which do not stop DNA replication and cell division, BrdUr inhibits the in vitro differentiation of melanocytes, cartilage cells, muscle cells, etc. It has no remarkable effects on already differentiated cells and on cell viability, but it specifically inhibits the expression of the *luxury* proteins: melanin, chondroproteins, contractile proteins in respectively, melanoblasts, chondroblasts and myoblasts. It has also been reported that treatment of very young chick embryos with BrdUr suppresses the formation of red blood cells and the synthesis of hemoglobin. However, treatment with BrdUr of fertilized or cleaving sea urchin, ascidian, amphibian, hen and mouse eggs has led to disappointing results: all one can see is a slowing down of development, due to cell injury and death. There is no specific inhibition of tissue differentiation when the drug is added to whole eggs or embryos.

Why BrdUr exerts highly specific effects on the expression of the luxury proteins remains obscure at the molecular level. Incorporation of the drug into replicating DNA obviously takes place, but it occurs randomly and we have only hypotheses for explaining the specific effects of the analogue on the genes which control the synthesis of the luxury protein. One possibility is that incorporation of BrdUr into DNA affects regulatory sequences controlling the activity of the structural genes coding for the luxury proteins. Another possibility is that certain chromosomal proteins bind specifically to the substituted DNA sequences: transcription of these sequences would then be shut off. Indirect effects on the cell membrane have also been suggested, but it is unlikely that they play a major role in the still mysterious BrdUr effect.

6.2.3 Effects of Phorbol Esters and Retinoids

The cancer problem lies outside the scope of the present book, but we cannot avoid mentioning it when cell differentiation is under discussion. Cancer is believed to be a multistep process (like embryogenesis!) leading to "mad" cells in which the regulatory mechanisms which control proliferation and differentiation in normal cells no longer function properly. All carcinogenic agents (physical, like X-rays; chemical, like the polycyclic hydrocarbons; viral, etc.) affect DNA. The cells in which DNA has been injured by a genotoxic agent will not necessarily become malignant: after *initiation* by a carcinogen, *promotion* by a co-carcinogen (promoter) is required as a second step. If, as shown by I. Berenblum almost 30 years ago, one applies locally a single dose of a carcinogen to the ear of a mouse, cancer will seldom develop, but if one rubs the ear of this mouse with croton oil, fast-growing tumors will form. Croton oil thus contains a *co-carcinogen* (promoter); the active principles of croton oil are *phorbol esters,* the more potent being 12-O-tetradecanoyl-phorbol-13-acetate (TPA).

At very low concentrations, TPA stimulates cell proliferation and, like BrdUr suppresses differentiation in cultures of myoblasts and chondroblasts. It is even able to induce the dedifferentiation of cultured myotubes and to disrupt selectively their myofibrils (Fig. 77). Differentiation of cultured epithelial cells is also suppressed by TPA: they fail to synthesize their marker protein, keratin. In induced MEL cells, TPA inhibits hemoglobin synthesis.

So far, the effects of TPA on whole eggs and embryos have been seldom studied: the most striking, in both sea urchin and amphibian eggs, is a dissociation of the superficial ectodermal cells. This suggests that the cell membrane is the initial site of action for TPA. There is now strong evidence for the view that TPA indeed acts on the *cell membrane:* in particular, cell-to-cell communication (i.e. the transfer of small molecular weight substances from a cell to its neighbors) is strongly decreased in TPA-treated cell cultures. Recently, it has been demonstrated that the cell membrane possesses specific receptors for phorbol esters. After binding to these receptors TPA activates a newly discovered membrane protein kinase (which requires phospholipids and Ca^{2+} for activity). It is now believed that this *protein kinase C* is the phorbol ester receptor. Its activation by TPA results in the phosphorylation of cytoplasmic proteins involved in the organization of the cytoskeleton and in protein synthesis.

It should be added that, as in the cases of cell hybridization and BrdUr treatment, the effects of TPA on cell differentiation are not always negative. For instance, TPA promotes the differentiation of a few human leukemic cell lines. We do not know why they make exceptions to the general rule; all one can say is that drugs like TPA and BrdUr exert "pleiotropic" effects, but this word masks our present ignorance.

It is interesting to note that the sites of action of BrdUr (DNA) and TPA (the cell membrane) are entirely different; yet the two drugs exert very similar effects on cell differentiation. Only more work will tell us whether TPA and BrdUr are ultimately acting on a single, still unknown, key step which would be absolutely required for cell differentiation. This step might perhaps turn out to be as simple as a change in the free calcium ion concentration or in the intracellular pH, since we know that such changes may have far-reaching consequences for the cell.

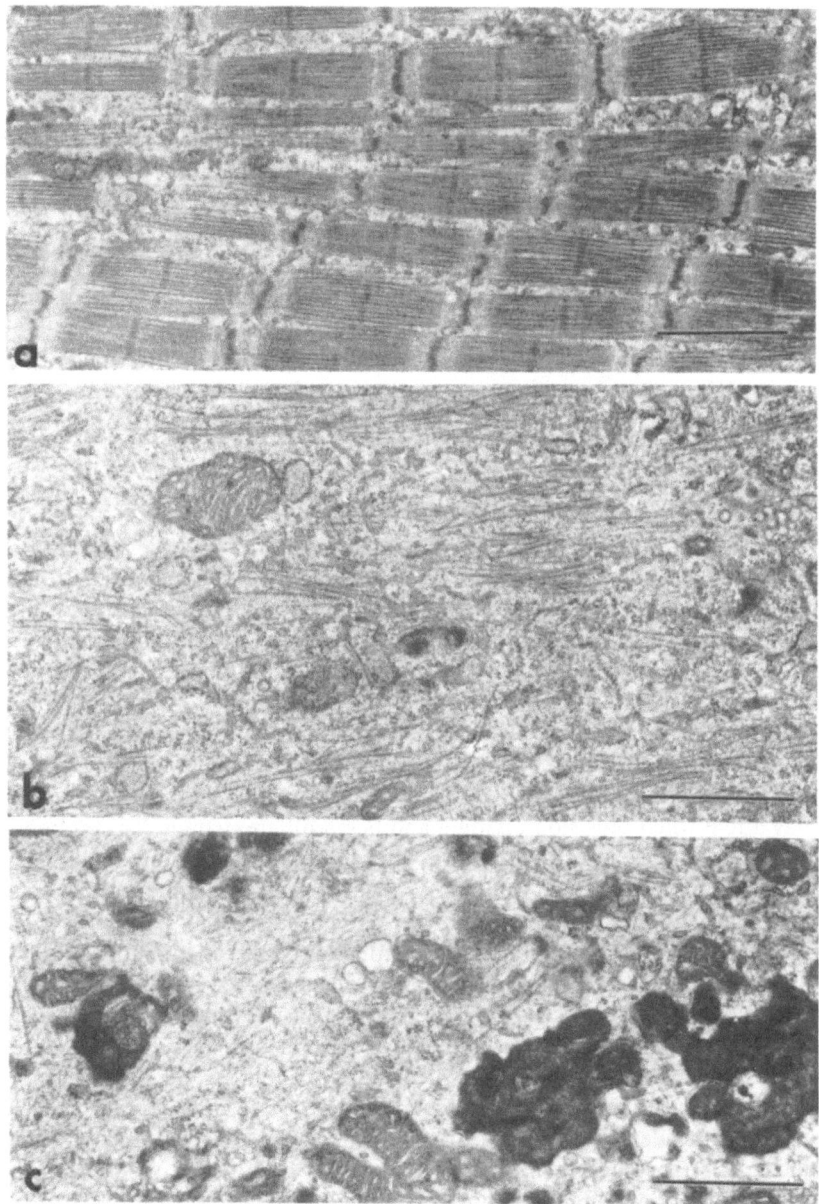

Fig. 77 a–c. Dedifferentiation of cultured myotubes by treatment with TPA: Electron micrographs of an untreated myotube (**a**) and of myotubes treated for 24 h (**b**) and 3 days (**c**). *Bar* 1 μm (Croop et al. 1982)

Derivatives of vitamin A (the *retinoids* related to the retina pigment retinene) often antagonize the promoting effects of phorbol esters on cancer cells. In certain malignant cell lines, *retinoic acid* (the most active of the retinoids) can induce the differentiation of undifferentiated cancer cells. These favorable effects on cell differentiation are linked to the fact that retinoic acid slows down the in vitro multiplication of malignant cells. The most striking effect of retinoic acid on cell differentiation has already been mentioned in this chapter: "nullipotent" (thus incapable of any kind of differentiation) embryonal carcinoma cells differentiate into endoderm if they are treated with retinoic acid. It is very likely that the action of the retinoids (which are liposoluble compounds) is mediated by a cytosolic retinoic acid binding protein (cRABP).

We do not know how the signals received by the cell, after treatment with either TPA or retinoic acid, are transmitted to chromatin where a few genes involved in cell differentiation must, in a selective way, be activated or shut down.

6.2.4 Other Inducers and Inhibitors of Cell Differentiation

It is likely that these signals are transmitted from the cell membrane to the nucleus, e.g. through *second messengers,* cyclic nucleotides or calcium ions. Increasing the cAMP content of the cells often slows down proliferation and favors differentiation. However, there are many exceptions to this rule. Similarly, there are numerous exceptions to the rule which states that a high cGMP content favors cell proliferation and prevents cell differentiation. As we have seen, a release of membrane-bound calcium ions triggers amphibian oocyte maturation and sea urchin egg fertilization. It is likely that changes in free Ca^{2+} and in internal pH play a role when cells stop dividing and begin differentiating. It is probable that cyclic nucleotides and calcium ions, when they trigger cell differentiation, activate protein kinases which can phosphorylate serine, threonine and tyrosine residues in proteins which are important for differentiation.

A few attempts to modify cell differentiation by the addition of so-called *conditioned media* to differentiating cells have been made. Conditioned media are culture media in which cells have been grown for a few hours or days. They contain substances which have been released or excreted in the surrounding medium by the growing cells. Conditioned media are valuable for the demonstration and isolation of substances which affect (positively or negatively) cell proliferation and of molecules involved in cell aggregation. However, no differentiation inducing substance has been isolated so far from conditioned media with the exception of a factor, present in the conditioned media of chondrocytes which stimulates the differentiation of these cells into cartilage. This factor, of unidentified chemical nature, stimulates the synthesis, by chondrocytes, of chondroproteins and collagen.

Popular today among the students of differentiation is *5-azacytidine,* which like BrdUr, is incorporated into replicating DNA molecules. The interesting property of this cytidine analogue is that it inhibits *DNA methylation.* As already mentioned, undermethylation of the cytidine residues in DNA molecules is a frequent – but not a constant – marker of "active" chromatin. In contrast to BrdUr, 5-azacytidine exerts *positive* effects on cell differentiation: for instance, it induces the differentiation of pre-adipocytes into adipocytes, of myoblasts into myotubes and of chondroblasts into cartilage cells. On the other hand, it has no effects (except sheer toxicity) on sea urchin and mouse eggs; unlike retinoic acid, it does not induce differentiation in nullipotent embryonal carcinoma cell lines. The most curious effect to date of 5-azacytidine is the *reactivation of the X chromosome* in females of the mammals (see Chap. IX).

Culture of a man-mouse hybrid cell line in the presence of 5-azacytidine is followed by reactivation of the X chromosome as shown by genetic analysis. However, only negative results have been obtained when normal diploid fibroblasts were treated with 5-azacytidine. Evidently, this substance is a promising tool for the analysis of cell differentiation; but more work is required before a close link between this process and DNA undermethylation can be accepted as a general fact.

7 Embryonic Differentiation and Cancer

Although carcinogenesis and embryogenesis are two different topics, the two processes have so much in common that a short comparison is not out of order here. Embryonic differentiation is the last step prior to senescence and death; embryogenesis is a multistep process which starts at gametogenesis. Malignant transformation, as mentioned before, is also a multistep process. It leads to the production of "transformed" cells, which can give rise to permanent cell lines. These cells are *immortal* and, in contrast to normal cells, never show signs of *senescence* (large size, low rate and finally arrest of DNA synthesis, modifications of the cytoskeleton, etc.). Instead, transformed cells go on dividing continuously. Although the initial lesion in the multistep carcinogenic process affects nuclear DNA, other cell constituents are also affected. In particular, the cell membrane properties are deeply modified in transformed cells. These cell surface changes allow transformed cells, in contrast to normal cells, to grow without attachment to a solid substratum (anchorage-independent growth). This, and the production by transformed cells of proteolytic enzymes, which can destroy the walls of the vessels, are responsible for the formation of the *metastases* which ultimately kill the host.

We have seen that our differentiated tissues derive from the proliferation of one or a few clones of cells; cancers also result from repeated division of a single malignant cell (clonal origin of cancers). All the cells which form a clone of normal, differentiated cells (for instance, all muscle or cartilage cells) look very much alike. This is not true for cancer cells: cells present in different metastases originating from the same tumor cell and even cells present in the same metastase display a striking *diversity*. This can be shown with immunological methods, which demonstrate a greater heterogeneity in malignant than in normal cells belonging to the same tissue. This heterogeneity might be due to genetic instability, leading to a high frequency of somatic mutations, in cancer cells. It is indeed well known that chromosomal abnormalities are frequent in cancer cells. However, the reasons why clones – in contrast to those of normal cells – of cancer cells undergo an increasing diversity when they expand by repeated cell divisions, remain mysterious.

It has often been suggested that cancer cells are more similar to embryonic cells than to adult cells. Cancer cells would result from the dedifferentiation of adult cells and this would allow them to proliferate repeatedly. This view is certainly an oversimplification of a more complex problem, but it cannot be disregarded. A classical example is the synthesis of two related proteins, α-fetoprotein and albumin, by liver cells. Embryonic liver synthesizes α-fetoprotein; around birth, the α-fetoprotein gene becomes silent; the albumin gene is switched on and albumin is synthesized during the whole lifetime. However, in liver tumors (hepatomas), the α-fetoprotein gene is reactivated and the albumin gene becomes inactive: thus, hepatomas synthesize an embryonic protein, α-fetoprotein. Another well-documented example is the frequent occurrence of *carcinoembryonic antigens* in tumors: these surface antigens are present in mammalian embryos and are no longer detectable in adults, but they are found in many tumors, in particular, in those of the colon. A particularly interesting case is that of the so-called F9 antigen of nullipotent malignant mouse embryocarcinoma (teratocarcinoma) cells. This antigen is shared with mouse sperm and morulae; it quickly disappears during development and it is absent from differentiated teratocarcinoma cells. The fact that teratocarcinomas may arise when early mouse embryos are grafted into an ectopic site and that malignant embryocarcinoma cells can undergo embryonic regulation, i.e. participate in normal development, after their introduction into a normal mouse blastocyst, show further that embryos and cancers must have much in common. The frequent reexpression of fetal genes in cancer cells has been called *retrodifferentiation*. However, there is an obvious difference between cancers and embryos: cancer cells proliferate and invade the host tissues in an anarchic way; embryos develop in a harmonious way because they are under the control of gradients, fields of differentiation and organizers. Positional information plays an essential role in normal morphogenesis; although it seems to be lost in neoplastic growth.

The origin of cancers has been the subject of extensive research in recent years. Since cancer is a multistep process (*initiation* by a genotoxic agent, *promotion* by agents which, like TPA induce a new phenotype, *proliferation* of the modified cells leading to a tumor by clonal expansion, formation and growth of *metastases*), it is impossible to pinpoint a single cause for all cancers. However, one of the factors which certainly plays an important role in malignant transformation, is the activation of cancer genes (*oncogenes*), which have been recently identified by molecular biologists. It has been known, for many years, that a number of viruses induce tumors (leukemias, sarcomas) in birds and mammals (including man). Some of these viruses contain DNA, others RNA; the latter are called *retroviruses* because they contain an enzyme, *reverse transcriptase,* which copies the viral RNA into the complementary DNA which is integrated into the host genome as a *provirus*. Proviral DNA contains an oncogene responsible for malignant transformation. When these oncogenes are expressed in virus-infected cells, transforming factors are produced. In many cases, the viral oncogene product is a protein kinase, which phosphorylates several cellular proteins and alters their conformation, leading to the transformed phenotype. Recently, it has been found that certain viral oncogenes increase the production of growth factors or of their receptors on the cell surface. One of the major surprises in this fast-moving field has been the demonstration that *normal cells* contain oncogenes (*cellular oncogenes*) very similar to the viral oncogenes. Activation of normal cellular oncogenes could thus lead to malignant transformation. Today we know two possible mechanisms for *oncogene activation*. One of them is *point mutation:* replacement of a single base by another in a cellular oncogene will lead to the production of the homologous transforming viral oncogene. The products of the two oncogenes will differ by a single amino acid substitution and this might have far-reaching consequences for the biological properties of the encoded proteins. The other mechanism for cellular oncogene activation is *translocation* of the gene from its normal site to a different site located on another chromosome. We have seen that chromosomal aberrations are frequent in tumors; among them are translocations in which a fragment of a chromosome is integrated in another chromosome. If a cellular oncogene is inserted in a chromosome region, which is favorable for transcription, it will be activated and the protein it is coding for will be expressed. Several examples in which cellular oncogenes are activated by translocation into the very active immunoglobulin locus (where antibodies are synthesized) are now known.

What is not known is the possible role of the cellular oncogenes in the normal cells which harbor them. These genes are silent in normal adult cells; but it is not excluded that they are active in the control of growth and differentiation in embryos. The fact that cellular oncogenes have been highly conserved during evolution and that they are present in *Drosophila* as well as in mice and men, indicate that they must have important functions. It will be an important task for the molecular embryologists of tomorrow to identify the products of the cellular oncogenes during development and to determine whether they play a role in proliferation and differentiation. If so, molecular embryology will benefit greatly from the recent progress made in cancer research.

8 The Future of Molecular Embryology

It would be absurd, for someone who has witnessed the stupendous progress made by molecular biology during the last 40 years to make predictions about the future development of molecular embryology. The discovery of a favorable material, which could be analyzed genetically, show good cell differentiation and lend itself to the preparation of in vitro systems, would be enough to make a big jump ahead. We need, among the eukaryotes, an organism which can be compared, in simplicity, with the bacteria and the viruses. Whether the ideal material for the study of cell differentiation really exists seems doubtful. All we can do now is to select systems that have advantages for the study of a particular problem and compare them. This is what we did, for example, when we compared *Acetabularia* and sea urchin eggs from the viewpoint of the morphogenetic and biochemical potentialities of nonnucleate cytoplasm. There is no doubt that the developing embryo, despite its complexity, remains the best material for the study of cell differentiation.

In the first edition of this book, we discussed a number of problems which seemed particularly important. We thought that their solution would be a turning point for molecular embryology, but this solution seemed so remote that they looked more like dreams than realities. Due to the astounding progress made by molecular biology, within little more than 10 years, these dreams have indeed become realities; molecular embryology has greatly profited – and will profit more and more in the future – from the availability of pure genes due to recombinant DNA technology.

Cloned genes (or the cDNAs of specific mRNAs) can be used for the detection, by in situ hybridization, of individual gene localization on metaphase chromosomes. We can now see, under the microscope, where specific genes (the hemoglobin genes, for instance) are located. Purified genes can be injected into the nucleus of a *Xenopus* oocyte where they will be accurately transcribed; these genes may be modified by chemical or enzymatic manipulation prior to their injection and this will tell us more and more about the control of transcription. Injection of mRNAs into the cytoplasm of *Xenopus* oocytes is now a standard method for the analysis of message translation. These oocytes are often a more convenient and efficient material than ribosomes isolated from bacteria, wheat germs or reticulocytes for those interested in the mechanisms of protein synthesis.

It is now possible, as we have seen, to inject pure genes into one of the pronuclei of a fertilized mouse egg and to obtain *transgenic* mice which express the injected gene. This seemed, 10 years ago, a very remote dream; the obtainment of transgenic mice is a major victory for molecular and experimental embryology. There is no doubt that we shall learn a good deal more about gene control in different tissues from this kind of experiment. It should be added, although the topic is outside the scope of this book, that the introduction of cloned, pure genes into somatic cells (fibroblasts, for instance) by addition of DNA to these cells is now a routine experiment: hundreds of genes have been introduced into cells by such *transfection* experiments and their fate (integration into the genome followed by replication, or degradation of the DNA molecules which have not been integrated) has been followed. In addition, somatic hybrids, in particular, those between mouse and human cells, have provided invaluable information about the localization of many genes on human and mouse chromosomes. Those who have witnessed the slowness of our progress in gene "mapping" on human chromosomes by classical genetic analysis of pedigrees are amazed to see that the mouse and human genomes might soon be known in as much detail as that of *Drosophila*.

We wondered, 12 years ago, whether the progress made in mammalian embryology would lead to the production of "babies in test tubes". We know that they now exist; artificial insemination of human eggs is now routinely performed in many hospitals; culture of the fertilized eggs and implantation until birth in their own or a foster mother has been successfully achieved without ill effects so far. We wish to point out again the ethical problems raised by experimental work on human eggs and sperm: by properly selecting eggs and sperm, one could try to obtain a higher proportion of Nobel Prize winners or football players. By separating blastomeres during cleavage prior to implantation, one could obtain, at will, identical twins. When one is able to separate the spermatozoa which possess the X or Y chromosome, it will be possible to shift, at will, the proportion of men and women in the world. Injection of genes into a pronucleus of a fertilized human egg might lead to transgenic men, a rather frightening prospect. All this would be Huxley's *Brave New World,* a novel which would be a perfect anticipation of a future society, if the key word was not missing: DNA.

The day will come when, thanks to the development of more and more sophisticated machines for DNA sequencing, we shall know the whole base sequence of the human, mouse, *Drosophila,* sea urchins, etc. genomes. This will not explain embryonic development, but we shall know completely the information received by the fertilized egg, which will finally lead to the formation of an adult belonging to the same species. DNA is a unidimensional molecule and living things are three-dimensional beings: to understand morphogenesis, we must know much more about the tridimensional structure of proteins and other large molecules. From the DNA base sequence, the primary structure of the encoded proteins may be deduced, but not the interactions between the tridimensional structures of these proteins with lipids or with other proteins. We doubt that, as was once humorously suggested by Sidney Brenner, if one fed all these data to a computer, a living mouse would come out of the machine (unless it was living in the computer).

Falling now from dreams to reality, we think that research in molecular embryology will go on very much as it does now: the experimental, reductionist approach, which has been so successful, will be followed for many years to come. Science progresses slowly: an experiment destroys a wrong idea, and this is progress; a new technology appears, often in a very different field, and this is another source of progress. All this leads to a better understanding of embryogenesis; perhaps the day will come when theoretical biology will be able to explain, in a mathematical formula, the unity and diversity paradox of the living world: but this is another dream and there are more urgent tasks ahead of us.

We should know more about gene regulation before we can understand embryonic differentiation; we should know much more about the molecular nature of the germinal localizations, of embryonic regulation and of inductions before we can indulge in theories. These are tasks for today and we do not doubt that outstanding progress will be made in the years to come. How long will it take before we understand how, at the molecular level, a hen egg becomes a chicken: a few decades, a few centuries? Perhaps we shall never know.

References

Foreword

Giudice G (1973) Developmental biology of the Sea Urchin embryo. Academic Press, New York

Chapter I

Brachet J (1945) Embryologie chimique. Masson, Paris, and Desoer, Liège
Holtfreter J (1935) Morphologisches Beeinflussung von Urodelenektoderm bei xeno-plastischer Transplantation. Arch Entw Org 133:367–426
Holtfreter J (1947) Neural induction in explants which have passed through a sublethal cy-tolysis. J Exp Zool 106:197–222
Needham J (1931) Chemical Embryology. Cambridge University Press
Needham J (1942) Biochemistry and Morphogenesis. Cambridge University Press and Bentley House, London
Waddington CH, Needham J, Brachet J (1936) Studies on the nature of the amphibian or-ganization centre. III. The activation of the evocator. Proc. Roy. Soc. London Ser. B 120, pp 173–207
Wehmeier E (1934) Versuche zur Analyse der Induktionsmittel bei der Medul-larplatenindunktion von Urodelen. Arch. Entwickmech. 132:384–423

Chapter II

Doolittle WF, Sapienza C (1980) Selfish genes, the phenotype paradigm and genome evo-lution. Nature 284:601–603
Klug A (1972) Assembly of tobacco mosaic virus. Federation Proceedings 31:30–42
Orgel LE, Crick FHC (1980) Selfish DNA: the ultimate parasite. Nature 284:604–607

Chapter III

Britten RJ, Davidson EH (1969) Gene regulation for higher Cells. A theory. Science 165:349–357
Brown DD (1981) Gene expression in Eukaryotes. Science 211:667–674
Clement AC (1952) Experimental studies on germinal localization in Ilyanassa. J Exp Zool 121:593–625
Davidson EH, Britten RJ (1979) Regulation of Gene Expression: Possible Role of Repeti-tive Sequences. Science 204:1052–1059
Holtzer H, Weintraub H, Mayne R, Mochan B (1972) The cell cycle, cell lineages, and cell differentiation. Curr Topics Dev Biol 7:229–256

Razin A, Riggs AD (1980) DNA methylation and gene function. Science 210:640–661
Scarano E (1969) Enzymatic modifications of DNA and embryonic differentiation. Ann Embryol Morphog Suppl 1:51–61
Weisbrod S (1982) Active chromatin. Nature 297:289–295

Chapter IV

Brachet J (1979) Oogenesis and maturation in amphibian oocytes. Endeavour 3:144–149
De Robertis EM, Gurdon JB (1977) Gene activation in somatic nuclei after injection into amphibian oocytes. Proc Natl Acad Sci USA 74:2470–2474
Korn LJ, Gurdon JB (1981) The reactivation of developmentally inert 5S genes in somatic nuclei injected into Xenopus oocytes. Nature (London) 289:461–465
Scheer U, Franke WW, Trendelenburg MF, Spring H (1976) Classification of loops of lampbrush chromosomes according to the arrangement of transcriptional complexes. J Cell Sci 22:503–519
Trendelenburg MF, Gurdon JB (1978) Transcription of cloned Xenopus ribosomal genes visualised after injection into oocyte nuclei. Nature (London) 276:292–294
Unwin PNT, Milligan RA (1982) A large particle associated with the perimeter of the nuclear pore complex. J Cell Biol 93:63–75
Wickens MP, Woo S, O'Malley BW, Gurdon JB (1980) Expression of a chicken chromosomal ovalbumin gene injected into frog oocyte nuclei. Nature (London) 285:628–634

Chapter V

Monroy A (1960) Incorporation of S^{35} methionine in the microsomes and soluble proteins during the early development of the sea urchin egg. Experientia 16:114
Van Assel S, Brachet J (1966) Formation de cytasters dans les oeufs de Batraciens sous l'action de l'eau lourde. J Embryol Exp Morph 15:143–151

Chapter VI

Brachet J (1933) Recherches sur la synthèse de l'acide thymonucléique pendant le développement de l'oeuf d'oursin. Arch Biol 44:519–576
Kriegstein HJ, Hogness DS (1974) Mechanism of DNA replication in Drosophila chromosomes, structure of replication forks and evidence for bidirectionality. Proc Natl Acad Sci USA 71:135–139
Newport JW, Kirschner M (1982) A major developmental transition in early Xenopus embryos I and II. Cell 37:675–686, 687–696
Parisi E, Filosa S, De Petrocellis B, Monroy A (1978) The pattern of cell division in the early development of the sea urchin Paracentrotus lividus. Dev Biol 65:38–49
Zeuthen E, Hamburger K (1972) Mitotic cycles in oxygen uptake and carbon dioxide output in the cleaving frog egg. Biol Bull 143:699–706

Chapter VII

Conklin EG (1905) Organization and cell lineage of the ascidian egg. J Acad Natl Sci (Philadelphia 13:1–119
Kalthoff H (1979) Analysis of a morphogenetic determinant in an insect embryo (Smittia sp. Chironomidae, Diptera). In: Subtelny S, Konigsberg JR (eds) Determinants of spatial Organization. Academic Press, pp 97–126
Lillie FR (1902) Differentiation without cleavage in the egg of the annelid Chaetopterus pergamentaceus Arch Entwicklungsmech Org 14:477–499

Chapter VIII

Ancel P, Vintemberger P (1948) Recherches sur le déterminisme de la symétrie bilatérale dans l'oeuf des Amphibiens. Biol Bull 31 (Suppl 1): 1–182
Sze L (1953) Respiration of the parts of the *Rana pipiens* gastrula. Physiol Zool 26: 212–223

Chapter IX

Fehilly CB, Willadsen SM, Tucker EM (1984) Interspecific chimaerism between sheep and goat. Nature (London) 307: 634–636
Mintz B (1962) Formation of genotypically mosaic mouse embryos. Am Zool 2: 432
Mintz B (1971) Allophenic mice of multi-embryo origin. In: Daniel JC (ed) Methods in Mammalian Embryology. W.H. Freeman and Co, pp 186–214
Pedersen RA, Spindle AI (1980) Role of the blastocoele microenvironment in early mouse embryo differentiation. Nature (London) 284: 550–552
Pincus G, Enzmann EG (1935) The comparative behavior of mammalian eggs in vivo and in vitro. I. The activation of ovarian eggs. J Exp Med 62: 665–675
Tarkowski AK (1961) Mouse chimaeras developed from fused eggs. Nature (London) 190: 857–860
Tarkowski AK, Wroblewska J (1967) Development of blastomeres of mouse eggs isolated at the 4- and 8-cell stage. J Embryol Exp Morphol 18: 155–180

Chapter X

Palmiter RD, Norstedt G, Gelinas RE, Hammer RE, Brinster RL (1983) Metallothionein-Human GH Fusion Genes stimulate Growth of Mice. Science 222: 809–814
Spring H, Grierson D, Hemleben V, Stöhr M, Krohne G, Stadler J, Franke WW (1978) DNA contents and numbers of nucleoli and pre-rRNA genes in nuclei of gametes and vegetative cells of *Acetabularia mediterranea*. Exptl Cell Res 114: 203–215

Chapter XI

Edelman GM (1983) Cell Adhesion Molecules. Science 219: 450–457
Eguchi G, Okada TS (1973) Differentiation of lens tissue from the progeny of chick retinal pigment cells cultured in vitro: a demonstration of a switch of cell types in clonal cell culture. Proc Natl Acad Sci USA 70: 1495–1499
Sato T (1930) Beiträge zur Analyse der Wolffschen Lensenregeneration. Roux's Arch 122, pp 451–493

Further Reading

Chapter I

Brachet J (1980) Cell differentiation yesterday and today. In: McKinnel et al. (eds) Results and problems in cell differentiation, vol 11. Differentiation and Neoplasia. Springer, Berlin Heidelberg New York, pp 1–7

Chapter II

Alberts B, Bray D, Lewis J, Raff M, Roberts K, Watson JD (1983) Molecular biology of the cell. Garland, New York

Brachet J (1941) La localisation des acides pentosenucléiques dans les tissus animaux et les oeufs d'amphibiens en voie de développement. Arch Biol 53:207–257

Brachet J (1985) Molecular cytology (2 vol) Academic Press, New York

Caspersson T (1941) Studien über den Eiweißumsatz der Zelle. Naturwissenschaften 29:33–43

Crick F (1971) General model for the chromosomes of higher organisms. Nature 234:25

Darnell Jr JE (1982) Variety in the level of gene control in eukaryotic cells. Nature 297:365–371

Feulgen R, Rossenbeck H (1924) Mikroskopisch-chemischer Nachweis einer Nukleinsäure vom Typ der Thymonukleinsäure und die darauf beruhende Elektive von Zellkernen in mikroskopischen Präparaten. Z Physiol Chem 135:203

Klug A (1983) From macromolecules to biological assemblies. Biosci Rep 3:395–430

McClintock B (1984) The significance of responses of the genome to challenge. Science 226:792–801

Nevins JR (1983) The pathway of eukaryotic mRNA formation. Annu Rev Biochem 52:441–466

Watson JD, Crick FA (1953) Structure for DNA. Nature 171:737

Weisbrod S (1982) Active Chromatin. Nature 297:289–295

Chapter III

Davidson EH (1976) Gene activity in early development. Academic Press, New York

DiBerardino MA, Hoffner NJ, Etkin LD (1984) Activation of dormant genes in specialized cells. Nature 224:946–952

Gurdon JB (1974) The control of gene expression in animal development. Clarendon, Oxford

Jacob F, Monod J (1963) In: Locke M (ed) Symposium on cytodifferentiation and macromolecular synthesis. Academic Press, New York

Monroy A, Moscona AA (1979) Introductory concepts in developmental biology. The University of Chicago Press, Chicago

Morgan TH (1934) Embryology and genetics. University Press, New York
Reverberi G (ed) (1971) Experimental embryology of marine and fresh water invertebrates. North-Holland, Amsterdam
Spemann H (1936) Experimentelle Beiträge zu einer Theorie der Entwicklung. Springer, Berlin Heidelberg New York
Taylor JH (1984) DNA methylation and cellular differentiation. Cell biology monograph, vol 11. Springer, Berlin Heidelberg New York

Chapter IV

Birnstiel M, Speirs J, Purdom I, Jones K, Loening UE (1968) Properties and composition of the isolated ribosomal DNA satellite of *Xenopus laevis*. Nature 219:454–463
Brachet J (1980) Induction of maturation in full-grown and vitellogenic amphibian oocytes by steroids and protein factors. In: Delrio G, Brachet J (eds) Steroids and their mechanism of action in nonmammalian vertebrates. Raven, New York
Brachet J (1985) Molecular cytology. Academic Press, New York
Brachet J, Malpoix P (1971) Macromolecular syntheses and nucleocytoplasmic interactions in early development. Adv Morphog 9:263–316
Brown DD, Dawid I (1968) Specific gene amplification in oocytes. Science 160:272–280
Callan HG (1979) Lampbrush chromosomes. Proc R Soc Lond B Biol Sci 214:417–448
Davidson EH (1976) Gene activity in early development, 2nd edn. Academic Press, New York
Huez G, Marbaix G, Hubert E, Leclercq M, Nudel U, Soreq H, Salomon R, Lebleu B, Revel M, Littauer UZ (1974) Role of polyadenylate segment in the translation of globin messenger RNA in *Xenopus* oocytes. Proc Natl Acad Sci USA 71:3143–3146
Lane CD (1983) The fate of genes, messengers and proteins introduced into Xenopus oocytes. Curr Top Dev Biol 18:89–117
Masui Y, Clarke HJ (1979) Oocyte maturation. Int Rev Cytol 57:185–282
Thomas C (1974) RNA metabolism in previtellogenic oocytes of Xenopus laevis. Dev Biol 39:191–197
Wellauer PK, David IB (1974) Secondary structure maps of ribosomal RNA and DNA. I Processing of *Xenopus laevis* ribosomal RNA and structure of single-stranded ribosomal DNA J Mol Biol 89:379–395

Chapter V

Brachet J (1960) The biochemistry of development. Pergamon, New York
Brachet J, De Petrocellis B (1981) The effects of aphidicolin, an inhibitor of DNA replication, on Sea Urchin development. Exp Cell Res 135:179–189
Davidson EH, Hough-Evans B, Britten RJ (1982) Molecular biology of the Sea urchin embryo. Science 217:17–26
Epel D (1978) Mechanisms of activation of sperm and egg during fertilization of Sea Urchin gametes. Curr Top Dev Biol 12:185–246
Jaffe LF (1983) Sources of calcium in egg activation: a review and hypothesis. Dev Biol 99:265–275
Lopo AC, Vacquier VD (1981) Gamete interaction in the Sea Urchin. A model for understanding the molecular details of animal fertilization. In: Mastroianni L Jr, Biggers JD (eds) Fertilization and embryonic development in vitro. Plenum, New York, pp 199–232
Monroy A, Moscona AA (1979) Introductory concepts in developmental biology. The University of Chicago Press, Chicago
Shapiro BM, Schackmann RW, Gabel CA (1981) Molecular approaches to the study of fertilization. Ann Rev Biochem 50:815–843

Chapter VI

Brachet J, De Petrocellis B (1981) The effects of aphidicolin, an inhibitor of DNA replication, on Sea Urchin development. Exp Cell Res 135:179–189

Brachet J, Malpoix P (1971) Macromolecular syntheses and nucleocytoplasmic interactions in early development. Adv Morphog 9:263–316

Callan HG (1972) Replication of the DNA in the chromosomes of eukaryotes. Proc R Soc Lond B Biol Sci 181:19–41

Davidson EH (1976) Gene activity in early development, 2nd edn. Academic Press, New York

Davidson EH, Hough-Evans B, Britten RJ (1982) Molecular biology of the Sea Urchin embryo. Science 217:17–26

Heby O (1981) Role of polyamines in the control of cell proliferation and differentiation. Differentiation 9:1–20

Kirschner M, Newport J, Gerhart J (1985) The timing of early developmental events in Xenopus. Trends Genet 1:41–47

Mazia D (1961) Mitosis and the physiology of cell division. In: Brachet J, Mirsky AE (eds) The cell, vol 3. Academic Press, New York, p 77–412

Chapter VII

Alexandre H, De Petrocellis B, Brachet J (1982) Studies on differentiation without cleavage in *Chaetopterus*. Requirement for a definite number of DNA replication cycles shown by aphidicolin pulses. Differentiation 22:132–135

Brachet J (1960) The biochemistry of development. Pergamon, New York

Davidson EH (1976) Gene activity in early development, 2nd edn. Academic Press, New York

Davidson EH, Hough-Evans B, Britten RJ (1982) Molecular biology of the Sea Urchin embryo. Science 217:17–26

Giudice G (1973) Developmental biology of the Sea Urchin embryo. Academic Press, New York

Harding K, Wedeen C, McGinnis W, Levine M (1985) Spatially regulated expression of homeotic genes in Drosophila. Science 229:1236–1242

McGinnis W, Garber RL, Wirz J, Kuroiwa A, Gehring WJ (1984) A homologous protein-coding sequence in Drosophila homeotic genes and its conservation in other metazoans. Cell 37:403–408

Reverberi G (ed) (1971) Experimental embryology of marine and fresh water invertebrates. North-Holland, Amsterdam

Satoh N (1984) Cell division cycles as the basis for timing mechanisms in early embryonic development of animals. In: Edmunds LN Jr (ed) Dekker, New York, pp 527–538

Whittaker JR (1979) Cytoplasmic determinants of tissue differentiation in the ascidian egg. In: Subtelny S, Konigsberg IR (eds) Determinants of Spatial Organisation. Academic Press, New York, pp 29–51

Chapter VIII

Brachet J (1960) The biochemistry of development. Pergamon, New York

Brachet J (1977) An old enigma: the grey crescent of amphibian eggs. Curr Top Dev Biol 11:133–186

Brachet J, Malpoix P (1971) Macromolecular syntheses and nucleocytoplasmic interactions in early development. Adv Morphog 9:263–316

Davidson EH (1976) Gene activity in early development, 2nd edn. Academic Press, New York

Dawid IB, Haynes SR, Jamrich M, Jonas E, Miyatani S, Sargent TD, Winkler JA (1985) Gene expression in *Xenopus* embryogenesis. J Embryol Exp Morphol Suppl. 89:113–124

Denis H (1974) Précis d'embryologie moléculaire. Presses Universitaires de France, Paris

Gerhart J, Ubbels G, Black S, Hara K, Kirschner M (1981) A reinvestigation of the role of the grey crescent in axis formation in *Xenopus laevis*. Nature 292:511–516

Kafiani C (1970) Genome transcription in fish development. Adv Morphog 8:209–284

Saxén L (1980) Neural induction: past, present and future. Curr Top Dev Biol 15:409–418

Spemann H (1936) Experimentelle Beiträge zu einer Theorie der Entwicklung. Springer, Berlin Heidelberg New York

Thiery JP, Duband JL, Tucker GC (1985) Cell migration in the vertebrate embryo: role of cell adhesion and tissue environment in pattern formation. Annu Rev Cell Biol 1:91–113

Tiedemann H (1968) Factors determining embryonic differentiation. J Cell Physiol Suppl 72, 1:129–144

Chapter IX

Epplen JT, Cellini A, Romero S, Ohno S (1983) An attempt to approach the molecular mechanisms of primary sex determination: W- and Y-chromosomal conserved simple repetitive DNA sequences and their differential expression in mRNA. J Exp Zool 228:305–312

Gardner RL (1985) Clonal analysis of early mammalian development. Philos Trans R Soc Lond B Biol Sci 312:163–178

Hyafil F, Morello D, Babinet C, Jacob F (1980) A cell surface glycoprotein involved in the compaction of embryonal carcinoma cells and cleavage stage embryos. Cell 21:927–934

Johnson MH, McConnell J, Van Blerkom J (1984) Programmed development in the mouse embryo. J Embryol Exp Morphol Suppl 83:197–231

Lyon MF (1961) Gene action in the X-chromosome of the mouse (*Mus musculus* L.). Nature 190:372–373

Mintz B (1974) Gene control of mammalian differentiation. Ann Rev Genet 8:411–470

Rossant J, Croy BA, Clark DA, Chapman VM (1983) Interspecific hybrids and chimeras in mice. J Exp Zool 228:271–286

Sherman MI (1979) Developmental biochemistry of preimplantation mammalian embryos. Ann Rev Biochem 48:443–470

Vandeberg JL (1983) Developmental aspects of X chromosome inactivation in eutherian and metatherian mammals. J Exp Zool 228:271–286

Wassarman PM (1983) Oogenesis: synthetic events in the developing mammalian egg. In: Mechanism and control of animal fertilization. Academic Press, New York, pp 1–54

Yanagimachi R (1981) Mechanisms of fertilization in mammals. In: Mastroianni L Jr, Biggers JD (eds) Fertilization and embryonic development in vitro. Plenum, New York, pp 81–182

Yoshida-Noro C, Suzuki N, Takeichi M (1984) Molecular nature of the calcium-dependent cell-cell adhesion system in mouse teratocarcinoma and embryonic cells studied with a monoclonal antibody. Dev. Biol 101:19–27

Chapter X

Brachet J (1960) The biochemistry of development. Pergamon, New York

Brachet J (1972) In: Bonotto S, Goutier R, Kirchman R, Maisin JR (eds) Biology and radiobiology of anucleate systems. Academic Press, New York

Brachet J (1981) A comparison of nucleocytoplasmic interactions in Acetabularia and in eggs. Fortschr Zool 26:15–34

Brachet J, Malpoix P (1971) Macromolecular syntheses and nucleocytoplasmic interactions in early development. Adv Morphog 9:263–316

Gordon JW (1983) Studies of foreign genes transmitted through the germ lines of transgenic mice. J Exp Zool 228:313–324

Harbers K, Kuehn M, Delins H, Jaenisch R (1984) Insertion of retrovirus into the first intron of α 1 (I) collagen gene leads to embryonic lethal mutation in mice. Proc Natl Acad Sci USA 81:1504–1508

Lüttke A, Bonotto S (1982) Chloroplasts and chloroplast DNA of *Acetabularia mediterranea:* facts and hypotheses. Int Rev Cytol 77:205–242

Magnusson T (1983) Genetic abnormalities and early mammalian development. In: Johnson MH (ed) Development in mammals, vol 5. Elsevier, Amsterdam, pp 209–249

Schweiger HG (1982) Interrelationship between chloroplasts and the nucleo-cytosol compartment in *Acetabularia.* In: Parthier B, Boulder D (eds) Encyclopedia of plant physiology, new series 14B: Nucleic acids and proteins in plants II. Springer, Berlin Heidelberg New York, pp 645–662

Vanden Driessche T (1980) Circadian rythmicity: general properties – as exemplified mainly by Acetabularia – and hypotheses on its cellular mechanism. Arch Biol 91:49–79

Chapter XI

Brachet J (1960) The biochemistry of development. Pergamon, New York

Brachet J (1985) Molecular cytology. Academic Press, New York

Bishop JM (1985) Trends in oncogenes. Trends Genet 1:245–249

Cairns J (1978) Cancer: science and society. Freeman, San Francisco

Davidson EH (1976) Gene activity in early development, 2nd edn. Academic Press, New York

Edelman GM (1985) Cell adhesion and the molecular processes of morphogenesis. Ann Rev Biochem 54:135–169

Guroff G (ed) (1983) Growth and maturation factors. Wiley-Interscience, New York

Harris H (1970) Cell fusion. The Dunham Lectures. Clarendon, Oxford

Holtzer H, Rubinstein N, Fellini S, Yeoh G, Chi J, Birnbaum J, Okayama M (1975) Lineages, quantal cell cycles, and the generation of cell diversity. Q Rev Biophys 8:523–557

Mintz B (1971) Clonal basis of mammalian differentiation. In: Control mechanisms of growth and differentiation. Symp Soc Exp Biol, vol 25. Cambridge University Press, London, pp 345–369

Monroy A, Rosati F (1979) Cell surface differentiations during early embryonic development. Curr Top Dev Biol 13:45–69

Ringertz N, Savage R (1976) Cell hybrids. Academic Press, New York

Author and Subject Index

Abortion 171
Acellular system 79
Acetabularia 155, 166, 169, 170, 204
– *crenulata* 155, 157
– *mediterranea* 155–157, 161
Acetate 80
Acetylcholinesterase 111, 112
β-N-acetylhexosaminidase 138
Achromosomal mitoses 89, 166
Acid phosphatase 64
Acini 192
Acquired tolerance 187
Acrosine 64, 138
Acrosome 64, 65, 67, 71
– membrane 67, 138
– reaction 67, 71, 137, 138
Actin 7, 39, 55, 68, 79, 90, 102, 104, 107, 111, 129, 158, 182, 193
– binding proteins 7
Actinomycin 6, 18, 50, 58, 60, 75, 77, 88, 102, 109, 111, 112, 117, 130, 142, 160, 185, 190, 193
Activation 58, 68, 72, 75, 76, 82, 106, 170, 172
Activator 35, 170
Actomyosin 8, 20, 94
Adenine 12
– phosphoribosyl transferase (APRT) 152
Adenosine 76
– diphosphate (ADP) 90, 91
– triphosphatase (ATPase) 9, 137
– triphosphate (ATP) 2, 9, 12, 20, 21, 59, 65, 91, 94, 104, 123, 124, 132, 160, 161, 174
Adenylate cyclase 12, 59
Adenylic acid 43, 76
Adipocytes 194, 201
Aggregation factors 117
Albumin 56, 137, 151, 202
Alcaline phosphatase 111, 112, 145
Alecithic egg 139
Alga 155, 160, 163
Allantois 139

Alleles 47, 150, 171, 177
Allophenic mice 144, 145, 154, 184
α-amanitin 18, 52, 102, 130, 141, 142
Amino acids 10–12, 17, 78, 115, 118, 175, 188, 203
β-amino acids 137
Ammonia 72, 75, 76, 80, 124, 131
Amnios 139
Amphibians 6, 28, 30–32, 36, 42, 47, 51, 52, 66, 71, 75, 78, 80, 86, 90, 91, 97, 99, 100–105, 121 ff., 140–142, 165, 166, 169, 172–176, 181, 184, 187–189, 191, 192, 197, 198, 200
Amphimixy 65, 67, 69, 70, 75, 80, 172
Amylase 191, 192
Anaerobiosis 91, 103, 118, 122, 174
Anaphase 61, 86, 91, 92, 94, 95, 104
Anastral mitoses 89
Anchorage-independent growth 201
Androgen 136
Androgenesis 171
Aneuploidy 89, 127, 171, 172
Animal pole 27, 30, 32, 39, 41, 58, 67, 87–89, 103, 105, 118, 123, 125, 126, 166
Animalization 117, 120
Animalizing substance, factor 117–120
Antennae 113, 114
Antibodies 14, 50, 55, 134, 135, 188, 203
Antigen 137, 153, 174, 175, 188, 202
Apex 163
Aphidicolin 75, 97, 107, 117, 148
Apolar cells 145
Arbacia 73, 120, 166, 173, 174
Archenteron 27, 116–118
Arginine 19, 65
ARISTOTELES 1
Artificial insimination 205
Ascaris 31
Ascidians 110 ff., 197
Ascites 171
Aster 67, 69, 70, 72, 83, 86, 89–92, 103, 166–169, 177
Astral rays 91

AUSTIN, C. R.   137
Autoradiography   6, 47, 97, 118, 160
AVERY, O. T.   12
Axolotl   33, 127, 173, 177
Axones   194
5-Azacytidine   37, 200, 201

"Babies in test tubes"   179, 205
Bacteria   10, 12, 13, 15, 18, 21, 54, 79, 95,
    110, 176, 192, 204
Bacterial transformation   12
Bacteriophages   13, 21–23, 25, 54, 65, 176
BALAKIER, H.   141
Barr body   152
Basal body   21, 65, 177
– plate   115, 116
Basement membrane   134, 151
Basic dyes   124
– proteins   64, 71, 175
BATAILLON, E.   82–85
BAUTZMANN, H.   3
BEADLE, G. W.   12
BERENBLUM, I.   198
BERG, W. E.   118
Bicarbonate   73
Bindin   67, 138
Biological clock   112, 143, 148
Birds   27, 123, 132, 202
BIRNSTIEL, M.   54
Bithorax complex   114, 115
Blastocoele   27, 87, 115, 134, 145, 148
Blastocyst   139, 141–143, 146, 147, 150,
    152, 154, 180
Blastocystogenesis   143, 148, 149
Blastomeres   27, 30, 86, 87, 89, 94, 100,
    104, 105, 109–111, 117–119, 139,
    145–147, 168, 176
Blastopore   122, 123, 125–127
Blastula   4, 32, 87–91, 97, 100–103, 112,
    115, 117, 118, 120, 125, 127, 133, 166,
    169, 172, 174–177, 187
Blastulation   139
Blood cells   83
Bone marrow   3, 125
BONOTTO, S.   164
BOVERI, T.   31
BRACHET, A.   70
BRACHET, J.   3, 4, 9, 12, 40, 42, 70, 95
Brain   90, 92, 95, 111, 127, 130, 196
BRENNER, S.   205
BRIGGS, R.   34, 35, 176
BRINSTER, R. L.   178, 179
BRITTEN, R.   36
Bromodeoxyuridine (BrdUr)   197, 200
BROWN, D. D.   36, 54

Bufo   51
Butyric Acid   70, 72, 83, 84, 167, 194

Cadherin   145
Calcium (Ca^{2+})   7, 20, 59, 61, 62, 67, 72,
    75, 84, 94, 105, 117, 119, 133, 138, 140,
    145, 175, 198, 200
CALLAN, H. G.   47, 49
Cancer   184, 197, 198, 201 ff.
– cells   195, 200, 202
– (tumor) viruses   13, 202
Cap (of mRNA)   14
– (of *Acetabularia*)   155, 159
– (of the oocyte chromatin)   47, 48, 54
Capacitation   137
Carbohydrates   74, 105, 118, 122, 123, 132,
    133, 164, 174
Carbonate   73
Carcinoembryonic antigens   202
Carcinogenesis   201
Carcinogens   198
Carnivora   139
Cartilage   182, 194, 197, 200, 201
Casein   41
CASPERSSON, T. R.   5, 9, 12
Catecholamines   137
Cauda epididymis   137
Cavitation   148
Cecidyomidae   31
Cell adhesion molecules (CAMs)   145,
    186, 187
– aggregation   200
– cycle   95–97, 101, 103, 129, 141, 142
– death (programmed –)   90, 176
– differentiation   7, 20, 28, 34, 37, 129,
    153, 173, 182 ff.
– division   9, 12, 13, 37, 38, 83, 84, 86, 88,
    94, 102, 103, 159, 183, 184, 188, 190, 196,
    202
– fusion   188, 195 ff.
– interactions   31, 145, 198
– junctions   7
– lineage   105, 151, 152
– membranes   2, 9, 67, 104, 132, 133, 145,
    158, 185 ff., 187, 188, 195–198, 200, 201
– movements, motility   27, 107, 117, 121,
    129, 132–134, 184
– organelles   8
– polarisation   145–147
– proliferation   198, 200, 202
– surface   59, 119
– wall   158, 161
Cells   2, 3, 5, 7, 10, 12, 20, 54, 80, 86, 90,
    92, 95, 97, 182, 196, 203
Cellularization   27, 105

Centrifugation   30, 78, 110, 131, 157, 166, 168, 169
Centrioles   38, 62, 65, 70, 72, 82, 83, 86, 89, 91, 92, 94, 169, 176
Centromeres   90, 92
Cerebratulus   81
Chaetopterus   81, 82, 105, 106, 112, 148
Chalones   187
CHAMBERS, R.   2
CHANG, M. C.   137
Chemotaxis   71, 133
Chemotherapy   94
Chiasmata   24, 47, 63
Chicken   4, 101, 115, 121, 123–125, 185, 191, 192, 196, 197
Chimaeras   150, 151, 153 ff.
Chimaeric organs   185
Chloramphenicol   18, 103, 161
Chloroplasts   35, 158, 160, 161, 162, 164, 165, 170
Cholesterol   74, 84
Cholinesterase   191
Chondroblasts   194, 197, 198, 201
Chondrocytes   194, 200
Chondrogenesis   194
Chondroitin sulfate   194
Chondroproteins   182, 194, 197, 200
Chorda   34, 121–125, 127, 173
Chordoblast   34, 123, 125, 131
Chordomesoblast   132
Chorion   36, 39
Chromatids   47
Chromatin   9, 19, 27, 36, 56, 62, 68, 83, 98–100, 119, 132, 152, 155, 175, 183
– activation   37, 183, 192, 201
– assembly factors   57
– condensation (factor)   141, 165
Chromatin-diminution   31
Chromomeres   47, 49, 50, 158
Chromosomal abnormalities   89, 90, 202, 203
Chromosome complement   171
– condensation   61, 62, 72, 103, 140, 141
– pairing   172
– replication   94, 104, 107
– unbalance   172
Chromosomes   4, 12, 13, 19, 24, 25, 31, 34, 39, 47 ff., 54, 57, 63, 64, 70, 72, 75, 82–84, 86, 89–94, 101, 159, 166, 169, 171, 172, 175, 194, 196, 203, 204
Chymotrypsin   192
Cilia   20, 21, 65, 102, 118, 124, 167, 169
Ciliation   106, 107, 115, 169
Circadian rhythms   160, 161, 164
Cistron   14

Cleavage   27, 32, 34, 60, 61, 67, 72, 74, 75, 86 ff., 107, 111, 115, 118, 122, 129, 132, 134, 139, 141, 148, 159, 166, 167, 169, 171, 172, 177, 180, 197
– arrest (cl) mutant   176
–, complete and equal   29, 86
– cycles   70, 83
–, partial   86, 121
–, spiral   107
–, unequal   28, 86
Clonal theory   184
Clones   184, 202
CO_2 73, 74, 81, 82, 84, 124, 131, 140
Co-carcinogens   198
Codon   14
Cognins   187
Colchicine   69, 83, 89, 94, 107, 133
Collagen   14, 43, 133, 179, 200
– gene   179
Commitment   143, 147, 183, 184, 193, 195
Compactin   117
Compaction   145, 148, 187
Competence   34, 124, 131
CONKLIN, E. G.   111
Contact guidance   133
Contractile proteins   20, 104, 129, 133, 158, 182, 193, 197
– ring   104, 105
Control DNA sequences   57
– of genetic activity 18, 20, 103, 142, 163, 170, 191, 204
Cordycepin   76
Corona radiata   138
Cortex   7, 39, 68, 74, 86, 104, 105, 107, 126, 127, 168
Cortical granules   39, 62, 69, 72, 84
– reaction   68–70, 72, 75, 78, 166
Corticosteroids   192
Cow   178
C-paradox   27
CRICK, F.   13
Crossing-over   25, 63, 64
Crystallin genes   189, 191
Crystallins   188, 190
Cumulus oophorus   138
Cyanide   91, 103, 122
Cybrids   196, 197
Cyclic adenosine monophosphate (cAMP)   12, 59–62, 140, 195, 200
– guanosine monophosphate (cGMP)   12, 200
Cyclin   103
Cycloheximide   18, 103, 127, 161, 162, 185
Cyclosis   158
Cynthia   111

Cysts   155, 158, 163, 165
Cytasters   69, 70, 89, 94, 169
Cytidine   200
Cytochalasin   90, 104, 105, 107, 111, 133, 148, 158, 171
Cytochemistry   12, 52, 88, 91, 106, 111, 128, 132, 158
Cytochrome oxidase   2, 91, 110, 123, 183
Cytochromes   3
Cytodifferentiation   145, 147
Cytokeratin   195
Cytolysis   84
Cytoplasm   5, 6, 7, 12, 18, 27, 30, 34, 36, 38, 39 ff., 46, 57, 59, 61, 62, 64, 69, 83, 84, 86, 89, 94, 101, 105, 106, 127, 140, 141, 155, 157, 158 ff., 160, 161, 166, 167, 174, 177, 180, 181, 196
Cytoplasmic control (of genetic activity)   112, 165, 175, 196, 197
– determinants   31, 38, 112, 113
– factors   57, 58, 88, 97, 110, 129, 164–166, 181, 197
Cytoplasts   196
Cytosine   12, 50, 141
Cytoskeleton   7, 8, 92, 133, 194, 198, 201
Cytostatic factor (CSF)   61, 94

DALCQ, A.   3, 130, 143, 145, 146
DAN, J. C.   91
DAVIDSON, E. H.   36
DAWID, I. B.   54
Decapacitation factor   137
Decidual reaction   143
Dedifferentiation   184, 190, 198, 202
Deep orange (dor) mutation   177
Dehydrogenases   2, 168, 183
Deletion   14, 37
Dendraster   174
Dendrites   194
DENIS, H.   129
Dentalium   109, 110
Deoxyadenosine   88
Deoxyribonuclease (DNase)   19, 37, 47
Deoxyribonucleic acid (see DNA)
Deoxyribonucleotides   97, 100, 141
Deoxyribose   12
DE ROBERTIS, E. M.   58
Determination   143, 145–147, 150
Dextrality   86
Diakinesis   24
Differentiated cells   183
Differentiation without cleavage   105, 106, 112, 148
Dimethylsulfoxide (DMSO)   194
Dinitrophenol   103

Dipeptidase   109
Diphteria toxin   127
Diploid   24, 25, 66, 83, 95, 158, 171, 172, 174
Diplotene   24, 39, 47
DISCHE, Z.   5
Discoglossus   176
Dispermy   69, 171
Dithiothreitol   69
DNA   5, 9, 12, 20, 25, 36, 47, 52, 56, 57, 65, 90, 99, 100, 109, 115, 158, 181, 183, 197
–, chloroplastic   38, 160–162, 164
–, chromosomal   39, 44, 54, 60, 100, 163
–, circular   19, 44, 52, 79, 164
–, complementary (cDNA)   57, 203, 204
–, cytoplasmic   25, 79, 100
–, extra-chromosomal   52
–, mitochondrial   18, 38, 60, 79, 100, 103, 141, 164, 170
–, nuclear   18, 27, 97, 101, 129, 161, 164, 165, 169, 201
–, repetitive   13, 27, 31, 36, 50, 76, 90
–, ribosomal (rDNA)   54, 60, 101, 102, 129
–, satellite   13, 90, 152
–, selfish   13, 27
–, viral   179
–, yolk platelet   100
DNA content   27, 39, 75, 79, 95, 107, 141, 162, 164, 170
– double helix   11, 13, 97
– endoreduplication   149
– exchanges   63
– methylation   37, 141, 142, 183, 200, 201
– modification   36
– recombination   204
– repair   13, 25, 64, 141
– replication   10, 13, 25, 27, 39, 44, 52, 54, 63, 72, 75, 82, 84, 87, 90, 95 ff., 100, 105, 141, 152, 165, 169, 175, 188, 192, 197, 200
– – cycles   112, 148
– sequencing   13, 14, 205
– synthesis   4, 60, 69, 75, 80, 89, 95, 97, 103, 104, 107, 109, 110, 112, 117, 129, 130, 133, 137, 148, 162, 163, 164 ff., 169, 175, 176, 188, 190, 193, 196
– synthesizing machinery   62, 97
DNA methylases   37, 141
DNA polymerase α   54, 56, 60, 75, 97, 102, 141, 148
– – β   141
DNA-binding proteins   20, 37, 183
Dominance   174
Donkey   153

Dorsal lip   123, 125–128, 130, 173
Dorsoventral organisation   125
Down's syndrome   172
DRIESCH, H.   30, 117, 119, 120, 145
Drosophila   13, 27, 31, 36, 97, 113–115,
    177, 179, 191, 203, 204
Dynein   65

Ear   130
Echinoderms   142, 174
Ectoblast   3, 27, 34, 122–124, 130–132,
    174, 185
Ectoderm   3, 34, 112, 116, 119, 122, 124,
    150, 151, 166, 180, 188, 198
Ectodermic growth factor (EGF)   188
Ectoplacental cone   149
EDELMAN, G. M.   186, 187
Eel   65
Egg water   71
EGUCHI, G.   190
Ejaculation   137
Elastin   133
Electron microscopy   1, 6, 39, 41, 45, 48,
    49, 52–54, 65, 68, 69, 88, 91, 98, 106, 110,
    132, 162, 164, 167, 168
Electrophoresis   74
Electrophysiology   195, 196
Elongation   17, 81
Embryogenesis   170, 201
Embryonal carcinoma cells (ECC)   180,
    195, 200, 201
Emetin   80
Endocytosis   9, 39–41, 188
Endoderm   112, 119, 124, 151, 200
Endoplasmic reticulum   8, 9, 43, 68, 104
Energy   8, 9, 17, 44, 60, 65, 81, 89, 90 ff.,
    103, 123, 140, 157, 160 ff., 165, 178, 182
Entelechy   30, 120
Entoblast   27, 122, 174, 185
Enucleation   35, 36, 60–62, 77–79, 83, 127,
    155 ff., 166 ff., 173, 175, 176, 196, 197
ENZMANN, E. G.   140
Enzymes   3, 8, 9, 11, 13, 19, 43, 44, 52, 60,
    64, 71, 74, 97, 161, 167, 175, 191, 196
EPEL, D.   71, 73
EPHRUSSI, B.   3, 12, 195, 196
Epiblast   151, 152
Epidermis   34, 124
Epididymal maturation   137
Epididymis   136, 137
Epigenesis   1, 110, 112, 145, 170
Epithelium   137, 198
Erythrocytes   195, 196
Erythropoiesis   183, 194
Esterases   175

Ether   90
Eucaryotes   15, 19, 20, 35, 57
Euchromatin   9
Evocating substance (evocator)   4, 124
Evolution   14, 15, 18, 203
Excretion   43, 56
Exocytosis   72, 148
Exogastrulation   32
Exons   14
Explants   4, 122, 124, 131, 132, 185
Extracellular matrix (ECM)   7, 133, 194
Extraembryonic ectoderm   149, 151
– endoderm   149, 151, 152, 195
– membranes, tissues   139, 143, 153
– mesoderm   151
Eye   90, 113, 114, 130, 132, 187, 189

F9 cells   195
– antigens   202
Fats   25
Fatty acids   124, 138
Fertility   137, 139
Fertilization   3, 4, 24, 25, 34, 37, 40, 58, 60,
    66 ff., 70, 79, 80, 84, 103, 110, 111, 135,
    148, 159, 181, 200
Fertilization acid   73
– membrane   67–69, 72, 78, 82, 83, 106,
    126, 137, 139, 140, 142, 167, 171, 172, 174
Fertilizin   71
Fetal genes   202
α-Fetoprotein   151, 195, 202
Fetus   139, 153, 178
FEULGEN, R.   5, 12
Fibroblast   188, 194, 196, 201
Fibronectin   133, 134, 151
Filopodia   132
Fishes   27, 121
Flagellum   20, 21, 25, 65, 137, 138, 167
Fluorescence microscopy   164
Fluorodeoxyuridine   88
Follicle   139, 140
– cells   40, 58
Follicular fluid   140
Formamide   94
Formation center   113, 143
Formyl methionine   17
Foster mother   143–145, 154, 178, 179, 205
Fragmented genes   14
FRANKE, W.   162
FRIEND, C.   194
Frog   4, 27, 30, 31, 67, 70, 73, 74, 79,
    81–84, 90, 91, 172, 173, 175
Fructose   65, 137
Furrow   27, 86, 90, 104 ff., 111, 133, 145, 167
Fusion-genes   178

GALL, J. G.   47, 49, 54
Gametes   24, 31, 136, 158, 164
Gametogenesis   24, 25, 27, 39 ff., 201
Gap junctions   145
GARDNER, R. L.   142, 143, 149
Gastropods   109
Gastrula   4–6, 29, 43, 101, 117, 122 ff.,
   127–129, 131, 166, 172–174, 185
Gastrulation   6, 29, 89, 100, 102, 104, 105,
   115–118, 121, 122, 128, 129, 132–134,
   166, 171, 173, 175, 176, 181, 188
Gene activation   6, 34–37, 55, 112, 129,
   166, 181, 194
– activity   19, 20, 34–36, 39, 163, 181, 183
– amplification   36, 52, 54, 162
– derepression   82, 102, 128, 130, 190
– dosage   196
– duplication   14
– expression   119, 141, 142, 151
– products   157, 177
– rearrangements   36, 114
– regulation   12, 20, 35, 57, 85, 162, 178,
   179, 183, 191
– repression   35, 175, 181
– therapy   180, 194
– transfer   115, 180
– translocation   36, 203
Genes   12–14, 19, 27, 31, 34, 36, 37, 49, 50,
   52, 55–58, 99, 110, 129, 135, 139, 151,
   173–175, 177, 178, 181–183, 189,
   192–194, 200, 204
–, extrachromosomal (nucleolar)   158
–, lethal   171
–, nuclear   162, 169
–, regulatory   35, 135, 161
–, ribosomal   52, 54, 162
–, structural   35, 192, 197
Genetic code   14, 15, 17
– continuity   10, 38
– factors   86
– message   17, 18
– recombination   25, 63, 64
Genitalia   136
Genome   36, 97, 142, 162, 165, 174, 203
–, chloroplastic   162, 165, 170
–, mitochondrial   170
Germ (line) cells   31, 133, 136, 152, 153,
   178, 179
– layers   153
– plasm   31
Germinal localizations   3, 31, 105, 106,
   110–112, 181, 205
– vesicle (GV)   5, 42, 44 ff., 58, 62, 67, 73,
   78, 89, 127, 166, 168, 176, 177

– – breakdown (GVBD)   58–62, 76, 103,
   140, 141
Germination   81
Giant trophoblastic cells   143, 150
Gills   83
Globin genes   178–180, 194
Glucose   91, 123, 137, 140, 188
– phosphate isomerase   150
Glutamine synthetase   191, 192
Glycogen   7, 11, 25, 39, 41, 74, 84, 87, 91,
   122–124
Glycogenolysis   122, 174
Glycolysis   74, 137, 174
Glycoproteins   14, 40, 67, 69, 120, 124,
   132, 133, 135, 137, 138, 140, 151, 158,
   175, 182, 187
Glycosaminoglycan   133
Glycosylation   43
Goat   154
Golgi bodies   9, 64, 158
Gonads   113, 135, 136
Gradients   3, 6, 169, 177, 202
–, animal-vegetal   6, 34, 123, 128, 131, 181
–, apico-basal   161, 164
–, cephalo-caudal (anteroposterior)   6, 34,
   125, 128, 131, 181
–, dorso-ventral   6, 34, 123, 128, 131, 181
–, mitotic activity   88
–, morphogenetic   6, 32, 34, 118, 128, 157,
   184
–, polarity   125
–, reduction   118
Grafts   157, 166, 180, 187, 202
Gravity   126
Grey crescent   31, 34, 125–127, 181
GROSS, P.   79
Growth hormone   188, 203
– – gene   178, 179
Guanine   12, 50
Guanylate cyclase   12
GURDON, J. B.   34, 35, 36, 37, 43, 58, 71,
   176, 182, 196
Gut   111
Gynogenesis   171, 172, 175
Gypsy mobile element   115

H-2 locus   178
HADORN, E.   114, 191
HAMBURGER, K.   92
HÄMMERLING, T.   157, 164, 165
Hamster   137, 139
Haploid syndrome   83, 171
Haploidy   24, 25, 64, 83, 158, 167,
   170–172, 174
HARRIS, H.   195

HARVEY, E. B.   167
Hatching   88, 102, 103, 106, 107, 112, 118, 167, 169, 173, 177
– enzyme   103, 169, 174
Head   32, 125, 187, 188
Heart   20
Heat shock   89, 166
Heavy bodies   76
– metals   71
– water (D_2O)   69, 70, 83, 89, 94, 169
HeLa cells   57, 61, 195
Hematopoietic cells   36, 37
Hematopoietin   194
Hemoglobin   10, 11, 14, 35–37, 43, 182, 183, 194, 197, 198
Hepatomas   202
Heredity   5, 12, 37, 178, 194
HERSHEY, A.   21
Heterochromatin   9, 152
Heterochromosomes   135
Heterogametic sex   135
Heterokaryon   196
Heterozygous   47, 176
Hexose-monophosphate   74
Histocompatibility antigens   178
Histone genes   99
– mRNA   43, 101, 102, 174
– synthesis   103, 109, 119, 165, 175
Histones   19, 20, 46, 57, 65, 79, 101, 141, 169, 174
–, core   19, 60, 62
–, Hl   60, 61
–, maternal   99, 100
HMG proteins   20
HOLTFRETER, J.   4, 123, 186
HOLTZER, H.   36, 182, 197
Homeo box   115
Homeotic gene   114
Homogametic sex   135
Homozygous   171, 176, 177, 179
Hormonal stimulation   17, 41, 58, 61, 90, 140, 192
Hormones   4, 9, 188, 192, 194
Horse   153
HÖRSTADIUS, S.   32, 117–119
"Housekeeping" proteins   183
HSU, Y. C.   148
HUEZ, G.   43
Human   115, 139, 171, 172, 179, 196, 202–204
HUXLEY, A.   205
Hyaline layer   69
Hyaloplasm   167
Hyaluronic acid   133
Hyaluronidase   64, 71, 138

Hybrid merogones   112, 175
Hybridomas   188
Hybrids, sexual   44, 67, 153, 171, 172 ff., 196
–, somatic   195–197, 201, 204
Hydrogen bonds   94
– peroxide (H_2O_2)   73
Hydrolases   64, 67
Hydrolytic enzymes   90, 191
Hydroxyurea   88
Δ^5, 3β-hydroxysteroid dehydrogenase (3β-HSD)   149
Hyperactivation   137
Hypermethylation   152
Hypertonic seawater   70, 72, 78, 82–84, 167
Hypoblast   151
Hypotaurine   137
Hypoxanthine phosphoribosyl transferase (HPRT)   152

IKEGAMI, S.   112
Ilyanassa   27, 28, 30, 109, 110, 168, 170
Imaginal disks   113, 114, 191
Immunoglobulins   56, 179, 188, 203
Immunological reactions, methods   153, 174, 187, 188 ff., 201
Implantation   139, 172, 177
Inducing substances   4, 5, 123 ff.
Induction (morphogenetic-)   3, 4, 33, 34, 105, 131, 184, 187, 205
Inductor   131
Initiation (of cancer)   198, 202
– codon   15, 16
– factors   17, 20, 99
– points   97–99
Inner cell mass (ICM)   139, 142, 143, 147, 150 ff., 152, 154, 180
Insects   27, 31, 90, 113 ff., 143
Insemination   75, 137, 153
Insertion   14
Insertional mutagenesis   179
Insulin   9, 41, 56, 59, 179, 194
Intercellular cement (matrix)   119, 120, 132
Intermediary pieces   121
Intermediate filaments   8
Interphase   91, 92, 95
Introns   14, 15, 27, 57
Invertebrates   51, 84, 106 ff., 139, 187
In vitro culture   139, 148, 151, 178, 184, 189, 193 ff.
Ionophores   72
Ions   41, 62
Iris   190
Iso-accepting tRNAs   17

JACOB, F. 18, 35, 191
JAENISCH, R. 179
Jelly coat 67, 71
JOHNSON, M. 145

KAFIANI, C. 121
KALTHOFF, H. 114
Karyocatalysis 83
Karyophilic proteins 46, 55, 177, 181
KCl 81, 107
Keratin 198
Kidney 34, 58, 121, 130, 132, 149, 185,
  187, 194
Kinetochore 90, 92, 94, 169
Kinetosome 21
KING, T. J. 34, 35
KIRSCHNER, M. 88, 101
KLUG, A. 22
KORN, L. J. 58

Lactate 140
Lactic acid 131
– dehydrogenase 162, 164
Lagging-strand template 98
Lamina 45, 60
Laminin 151, 195
Lampbrush chromosomes 47, 55, 58, 80,
  158, 159
Lanthanum chloride 59
Larva 187
Lateral plates 121
Laying 74
Leading-strand template 98
Lecithinase 74, 84
Lectin 105, 117, 133, 145, 175
Lens 34, 132, 188–190
Lentoid 190, 191
Leptotene 24
Lethal 89, 115, 128, 171–173
– hybrids 172 ff.
LETOURNEAU, M. J. 141
Leukemia 202
– virus 194
Leukemic cells 194, 198
LILLIE, F. R. 71, 109
Limb bud 90, 148
LINDAHL, P. E. 118
Linker DNA 19
Lion 153
Lipid droplets 167, 194
Lipids 3, 39, 41, 63, 74, 109, 137
Lipovitellin 41
Lithium chloride (LiCl) 117, 118, 124
Liver 3, 35, 41, 79, 124, 178, 187, 192, 196,
  202

Loach 121
Lobeless embryos 28, 109
Locus 14
LOEB, J. 72, 82, 85, 167
Luteinizing hormone (LH) 140
LÜTTKE, A. 164
"Luxury" proteins 182, 184, 197
Lymnaea 86, 109
Lymphocytes 36
LYON, M. 152
Lysine 17, 19
Lysolecithins 74
Lysophospholipids 138
Lysosomes 9, 64, 90
Lysyl-tRNA 17

Macromeres 88, 119, 167
Mactra 81
Magnesium (Mg^{2+}) 59, 119, 175
Maize 13
Malic dehydrogenase 162, 164
Malignancy 180, 184, 196, 197, 201
Malignant cells 201
– transformation 201, 203
Mammals 37, 62, 64, 65, 86, 89, 97, 101,
  102, 121, 123, 128, 135 ff., 171, 178, 187,
  201, 202
Manganese (Mn^{2+}) 59
MARBAIX, G. 43
Masked mRNA 78–81, 102, 107, 143, 163
Maternal mRNA 38, 76, 79–81, 101–103,
  109, 110, 113, 119, 129, 134, 140, 142,
  168–170, 176
Maturation of mRNA 14, 37
– of oocytes 37, 39, 55, 57, 58 ff., 66, 69–71,
  75, 78, 80, 81, 84, 89, 99, 103, 106, 127,
  139, 140 ff., 165, 200
– promoting factor (MPF) 61, 62, 103,
  105, 140, 165
– of rRNA 54
MAZIA, D. 91, 93, 94
Meiosis 24, 26, 39, 47, 58, 61–64, 67, 136,
  140, 141, 152, 158, 172
Meiotic inducing substance (MIS) 136
– preventing substance (MPS) 136
Melanin 190, 194, 196, 197
Melanoblasts 197
Melanocytes 194, 196, 197
Melanosomes 42
Membrane depolarization 72
– fusion 67, 138
– potential 72
Membranes (see cell and plasma membrane)
–, mitochondrial 18

MENDEL, G.   128
Mercaptoethanol   94, 132
Merogony   112
Mesenchyme   115, 116, 192
– blastula   102, 115, 117
Mesoblast   27, 122, 123, 174, 185
Mesoderm   121, 124, 125, 134, 151, 153, 182, 192
Mesodermalizing substance (protein)   4, 5, 132
Mesomeres   112, 167
Metabolism   12, 74, 84, 118, 122, 173–175, 191
Metallothionein (gene)   178, 179
Metamorphosis   90
Metaphase   61, 67, 86, 91, 94, 95, 104, 177, 204
Metazoans   135
Methionine   17, 77
Methotrexate   36
Methylation   37
Methylene blue   4, 124
7-methyl-guanosine 5′triphosphate (7 mGppp)   15
Microcephaly   30, 127, 130, 131, 171
Microfilaments   7, 8, 39, 68, 104, 105, 107, 111, 133, 145
Microinjection   43, 46, 55–57, 60, 62, 71, 75, 77, 86, 97, 100, 102, 105, 113, 127, 130, 150, 176–178, 196
Micromanipulation   139, 141
Micromeres   32, 88, 100, 116–119, 167
Microtubule organizer center (MTOC)   83, 92, 169, 177
Microtubule-associated proteins (MAPs)   92
Microtubules   7, 8, 61, 65, 69, 70, 72, 83, 91–94, 107, 133, 145, 177
Microvilli   39, 40, 62, 68, 145
Midblastula transition   88, 101, 129
Midge   113
Mid-piece   65
MINTZ, B.   144, 145
Misgurnus fossilis   121
Mitochondria   8, 17, 18, 20, 25, 38, 44, 65, 81, 103, 106, 109, 110, 123, 141, 167, 168, 170
Mitosis   27, 67, 83, 86–89, 92, 95, 103–105, 117, 149
Mitotic abnormalities   127
– activity   97, 99, 100, 130, 187, 190, 192
– apparatus   91 ff., 104, 105, 132, 133, 166
– index   88, 95
Mobile genetic elements   13, 14, 36, 115

Molecular clock   107
– differentiation   142, 148
– hybridization   49, 54, 129, 152, 175, 204
Mollusks   81, 86, 106, 168, 170
Monaster   69, 82, 106, 107
Monasterial cycle   84
Monoclonal antibodies   188
MONOD, J.   18, 35, 191
Monosomy   172
Monotremata   139
MONROY, A.   77, 79, 80, 186
MORGAN, T. H.   34, 110, 181
Morphogenesis   30, 31, 125, 127, 128, 130 ff., 132, 135, 142, 143, 155, 161–163, 165, 169, 171, 202, 205
Morphogenetic fields   3
– movements   105, 123, 134, 187
– potentialities   165, 204
– substances   157, 158, 159 ff., 161, 163–165
Morula   87, 88, 117, 119, 120, 141, 147, 152, 154, 167, 171, 175, 202
"Mosaic" eggs   30, 31, 105, 181
Mosaic embryos   171
– mice   144, 154
MOSCONA, A.   185, 186
Motility factor   137
Mouse   27, 86, 97, 101, 112, 115, 139, 141, 148, 171, 177, 178, 185, 187, 197, 198, 201–204
Mucopolysaccharides   69, 71
Müllerian ducts   135
– Inhibitory Factor (MIF)   135
Multipolar mitoses   171
Mural trophectoderm   149
Murine erythroleukemic cells (MEL)   194, 198
Mus caroli   152, 154
Mus musculus   152, 154
Muscle   8, 20, 34, 94, 121, 124, 129, 130, 184, 193
– cells   37, 110, 111, 182, 201
Mutagenesis   57
Mutants   176 ff.
Mutation   11, 13, 37, 57, 90, 114, 115, 128, 171, 177, 203
Myoblast   193, 197, 198, 201
Myocyte   184
Myofibril   182, 198
Myogenesis   183, 193, 194
Myoplasm   110, 111
Myosin   7, 158, 182, 193
Myotube   193, 194, 198, 201
Mytilus   168

NaCl   124
Na$^+$/H$^+$ exchange   73
NAD-kinase   73
NAD-NADP   73
Narcotics   20, 105
NEEDHAM, J.   1, 3, 4
Nematodes   31
Neoplastic growth   202
Nerve cells (neurones)   3, 28, 95, 124, 129, 187, 191, 194–196
– fiber   191
– growth factor (NGF)   194
Nervous system   3, 4, 28, 31, 34, 121–127, 130, 173
Neural induction   34, 122–127, 130, 132, 173, 181, 191
– plate   132
– tube   129, 132
Neuralization   124
Neuralizing substance, factor   4, 5, 124, 125, 130, 132
Neurites   194, 195, 197
Neuroblast   122, 131
Neuroblastoma   191, 195, 197
Neuroglioma   195
Neuro-teratocarcinoma stem cells   195
Neurula   6, 44, 121, 122 ff., 132, 188
Neurulation   100, 105, 121, 122, 128, 132, 133, 171
Neutral lipids   178
– red   4, 124
NEWPORT, J.   88, 101
Newt   4
No cleavage (nc) mutant   177
Nocodazole   94
Nonsense codon   17
Notochord   121
Nuclear control (of gene expression)   164, 165, 170, 171, 197
– division   90
– genes   38, 112
– matrix   9, 20, 55
– membrane   9, 19, 45 ff., 55, 56, 58, 60, 61, 68, 69, 91
– pores   9, 45, 60
– sap (nucleoplasm)   55 ff., 57, 69, 89, 141, 158, 176
– transplantation   34, 36, 37, 58, 173, 177, 181, 182
Nuclease   100, 119
Nucleic acids   4, 10–12, 21, 31, 55, 103, 123, 127, 174
Nucleocytoplasmic interactions   155 ff., 165
– ratio   27, 88, 105, 129

Nucleolar organizer   52, 54, 58, 60, 101, 102, 128, 129, 176, 196
Nucleolus   6, 9, 12, 19, 47, 51 ff., 58, 60, 88, 89, 101, 128, 155, 162, 165, 190, 196
Nucleolusless (nu-o) (anucleolate) mutants   128, 176
Nucleoplasmin   56, 57
Nucleoproteins   10, 127
Nucleosome   19, 20, 57, 99, 100, 129, 169
– assembly factor   56
Nucleotide   12, 124
Nucleus   3, 5, 6, 7, 9, 12, 19, 20, 27, 31, 34, 37, 38, 44, 46, 57, 58, 62, 64, 71, 79, 89–91, 97, 104, 107, 113, 127, 130, 132, 141, 152, 155, 158 ff., 161, 167, 171 ff., 180, 192, 195–197, 200
Nullipotent   27, 200
Nullo-x condition   177

OKADA, T. S.   190
Okazaki fragments   98
Oligosyndactylia (Os) mutant   177
Oncogene   202, 203
Oocyte   4, 6, 24, 27, 36, 39, 41, 42, 47, 52, 56–62, 68, 72, 73, 77, 78, 80, 128, 138–140, 166, 176, 177, 181, 204
Oogenesis   25, 36, 38, 39 ff., 63, 75, 79, 80, 86, 101, 177, 180, 191
Oogonium   24, 39
Ootide   24, 67
Operator   18
Operon   19, 192
Optic cup   132, 188
Organizer   3, 4, 34, 122, 124, 130, 166, 187, 202
Organogenesis   28, 121, 129, 130
Organomercurials   59
Oriented migration   133
Ornithine   104
– decarboxylase (ODC)   104, 163
Ova deficient (o) mutant   176, 177
Ovalbumin   56, 57, 192
Ovaries   31, 39, 54, 135, 136, 180
Oviduct   81, 135, 192
Ovocylinder   151
Ovoperoxidase   73
Ovulation   140
Oxidations   8, 20, 81, 122, 123, 175
Oxidative phosphorylations   65, 81, 91, 183
Oxygen consumption   3, 4, 60, 72, 73, 81, 84, 90, 91, 118, 122, 123, 160, 168, 173, 174

Pachytene   24, 47, 52, 64
Pancreas   4, 34, 132

Pancreatic enzymes   192
Paracentrotus   120, 173, 174
PARDUE, M. L.   54
Parietal endoderm   151, 195
PARISI, E.   96
Parthenogenesis   72, 82–84, 167, 171, 172, 174, 175
Parthenogenetic activation   4, 70, 82 ff., 166, 167, 169
PEDERSEN, R. A.   147
Penis   136
Peptide   119
– bond   10
Pericentriolar cloud   92
Peritoneal cavity   121
Perivitelline space   67
Permeability   46, 56, 77, 84, 175, 184
pH   3, 4, 67, 72, 73, 75, 80–82, 85, 124, 188, 200
Phage (see Bacteriophage)
Phase-contrast microscopy   49
Phenotype   196, 197, 202
Phorbol esters   197 ff., 200
Phosphatases   161, 191
Phosphate   175
Phosphatides   74
Phosphoglycerate kinase (PGK)   153
Phospholipase   138
Phospholipids   65, 84, 198
Phosphoproteins   41, 52, 61
Phosphorylations   81, 91
Phosvitin   41
Photosynthesis   157, 158, 160, 161, 165
Pig   178
Pigment (granules)   40, 126, 167, 190, 196
– cell   111, 191
PINCUS, G.   140
Pinocytosis   132
Pituitary gland   58, 140, 178
Placenta   139
Plants   10, 38, 63, 81, 89
Plasma membrane   9, 39, 40, 105, 132, 137, 138, 145, 175
Plasmagenes   38
Plasmids   57
Plasminogen activator   149, 151, 195
Plasms   110, 111
Pleurodeles   51, 58
Pluricentric mitoses   89, 127
Pluripotentiality   142, 143, 151, 180, 195
Pluteus   79, 82, 116, 117, 119, 167, 174, 175, 191
Pneumococci   12
Podophyllin   94

Polar body   24, 58, 61, 67, 73
– cells   145, 146
– lobe   30, 107, 109, 168, 170
– trophectoderm   149
Polarity   6, 27, 39, 41, 158, 181
Pole plasm   31, 113
Polyadenylation   76, 169
Polyadenylic (polyA) tract, tail, sequence   43, 56, 76, 101, 163, 169
Polyamines   103, 104, 163
Polychaete   81, 105
Polycyclic hydrocarbons   198
Polyethylene glycol   195
Polynucleotides   12
Polypeptide   10, 11, 17, 81, 142, 168, 187
Polyploidy   83, 89, 107, 143, 149
Polyribosomes (polysomes)   15, 17, 18, 20, 23, 78, 79, 81, 101, 128, 161, 168
Polysaccharides   7, 158, 161
Polyspermy   69–72, 75, 171, 172
–, prevention of   73
Polyuridylic acid   80
Positional information   145, 184, 202
Postimplantation   148, 171
Post-transcriptional control   37, 60, 82, 142
Potassium (K^+)   7, 59
Potentialities   3, 30–32, 78, 143, 149, 151, 195
Preadipocytes   194, 201
Precentriole   69, 169
Predetermination   145
Preformation   1, 110, 112, 170
Pregnancy   153, 172, 179
Preimplantation   139, 141 ff.
Previtellogenic oocytes   39, 42–44, 50, 51, 55
Primary mesenchyme   115, 116
– transcript   37
Primase   98
Primer   98
Primitive endoderm   151
Processing   37, 54, 56, 57
Procollagen (type-IV)   151, 195
Progesterone   59–62, 72, 78, 165, 192
Prokaryotes   19, 35
Promotion (of cancer)   198, 202
Promotor/regulator (of genes)   14, 178
Pronephros   125, 187
Pronucleus   67, 69, 70, 71, 75, 77, 78, 168, 170, 171, 178, 180, 204
Propanolol   59
Prophase   24, 86, 91, 159
Protamines   65
Proteases   74, 80, 84, 103, 163, 191

Proteins   5, 9–11, 13, 23, 39, 40, 46, 55, 63, 69, 74, 84, 87, 88, 91, 92, 101, 103, 118, 124, 129, 158, 162, 170, 188, 196
–, chloroplastic   161, 162, 164
–, chromosomal   19
–, cytoplasmic   71, 103
–, membrane   104
–, mitochondrial   79, 103, 161, 170
–, nucleolar   52
–, ribosomal   15, 52, 54, 74, 80, 101, 128, 129, 180
–, secretory   9, 43, 56
–, yolk   25, 39–41
Protein glycosylation   117, 133, 145
– kinases   12, 60, 140, 198, 200, 203
– phosphatase   61
– phosphorylation   12, 45, 60–62, 80, 81, 166, 168, 198, 200, 203
– subunits   10, 11, 92
– sulfation   117
– synthesis   3, 5, 6, 7, 9, 12–16, 18, 20, 42, 44, 45, 58, 60–62, 77 ff., 82, 83, 85, 89, 90, 101, 103 ff., 109, 113, 117, 118, 123, 127, 130, 131, 137, 140–142, 161 ff., 165, 166, 174 ff., 176, 192
– –, control of   10, 81, 85, 162
Proteinase   138
Proteoglycans   133, 194
Proteolysis   74, 81
Proteolytic enzymes   169, 201
Protons (H⁺)   72
Protoplasmic streaming   164
Pseudocleavage   106
Pseudogastrulation   106, 107
Pseudogenes   14
Pseudohybrids   172, 175
Pseudopodia   132
Pseudopregnancy   178
Puberty   136
Purines   12
Puromycine   6, 18, 80, 103, 112, 117, 131, 161, 162
Putrescine   103, 104, 163
Pycnosis   90
Pyrimidines   12
Pyruvate   140

Quantal mitosis   112, 183
– replication cycle   148

Rabbit   101, 139, 141, 142, 178
Rana catesbeiana   173
Rana pipiens   35, 173
Rana sylvatica   173
RAPKINE, L.   3

Receptors   9, 40, 133, 138, 140, 145, 158, 188, 192, 198, 203
Recessive   196
Recombination (see: genetic recombination)
Red blood cells   7, 83, 121, 125, 182, 183, 194, 195, 197
– pigment   167
Redox potential   3
Reducing center   118
Regeneration   157, 160, 161
Regulation
–, embryonic   3, 30, 32, 105, 117–119, 145, 181, 205
–, metabolic   73, 198
Regulative eggs   30, 105, 106, 145
Regulatory DNA sequences   14, 197
– factor   163
Reichert membrane   151
REISS, P.   3
Repeated DNA sequences   13, 27, 36, 50, 163
Replication (see DNA replication)
– cycles   88, 107, 112, 142, 148
Replicons   97, 98
Repressor   18, 19, 35, 101, 170, 173
Reptiles   121
Respiration   4, 17, 71, 72, 82–84, 91, 110, 118, 122 ff., 167, 168, 173
Respiratory chain   91
– enzymes   18, 44, 65, 168
– quotient (R. Q.)   74, 122
Reticulocytes   204
Retina   185, 189–192
Retinene   200
Retinoic acid   195, 200
– – binding protein (cRABP)   200
Retinoids   197 ff., 200
Retrodifferentiation   202
Retrovirus   141, 202
REVERBERI, G.   106, 110
Reverse transcriptase   57, 202
Revitalization   172, 173
Rhizoid   155, 163
Ribonuclease   5, 20, 130, 160
Ribonucleic acid (see RNA)
Ribonucleoprotein (particles)   31, 38, 42, 46, 50, 55, 76, 80, 88, 113, 163
Ribonucleotides   97
Ribonucleotide reductase   97
Ribose   12
Ribosomes
–, chloroplastic   162
–, cytoplasmic   6, 9, 15, 17, 18, 20, 23, 41–44, 52, 76–81, 91, 92, 101, 109, 124,

125, 128, 129, 140, 142, 162–164, 167, 168, 170, 176, 180
–, mitochondrial   162
Rifampicine   18, 203
RIGGS, A. D.   37
RNA   9, 12–14, 18, 23, 39, 44, 46, 47, 60, 63, 76, 88, 91, 98, 128, 158
–, chloroplastic   162
–, cytoplasmic   64
–, heterogeneous   56, 129, 163
–, messenger (mRNA)   14, 17–21, 37, 42–44, 46, 47, 56, 57, 76–78, 80, 100–102, 111, 119, 125, 129, 140, 142, 159 ff., 161–164, 168–170, 174, 180, 190, 192, 193
–, mitochondrial   44, 75, 79, 103, 170
–, nuclear   14, 36, 76, 160
–, nucleolar   52, 128
–, ribosomal (rRNA)   15, 18, 36, 42, 46, 52, 56, 60, 76, 100–102, 128, 129, 140, 142, 158, 163, 165, 170, 172, 176, 180, 190
–, – 28S + 18S   15, 19, 42, 46, 51, 52, 54, 128, 162, 163
–, – 5,8S   17, 54
–, – 5S   17, 19, 42–44, 46, 54, 58, 101, 129, 166
–, small nuclear (snRNAs)   46, 56, 129
–, 7S   129
–, transfer (tRNA)   17–20, 42, 43, 46, 56, 57, 101, 128, 129
–, viral   23, 203
RNA content   128, 142, 155, 161, 162
– distribution   6, 109, 119, 125 ff., 140, 169
– gradient   6, 118
– migration   160
– synthesis   6, 18, 42, 43, 47, 50, 58, 60, 65, 69, 75 ff., 76, 79, 88–90, 100 ff., 106, 109, 112, 117, 125 ff., 130 ff., 137, 140–142, 158, 162 ff., 172, 176, 177, 180, 190, 192, 196
RNA polymerase I   52, 54
– II   14, 50, 115, 141
– III   55
RNA polymerases   19, 20, 47, 56, 102, 140, 192
Rodents   140
Rolling circles   54
RUNNSTRÖM, J.   3, 73, 118, 119
RUTTER, W.   192

Sarcoma   202
SATO, T.   189
SATOH, N.   112
SCARANO, E.   36, 37
SCHWEIGER, H. G.   161, 162

Scrotum   136
Sea urchins   5, 25, 27, 30, 32, 37, 57, 67–84, 87–91, 94–106, 115 ff., 125, 129, 138, 140, 141, 166, 167 ff., 169, 170, 172, 174, 175, 187, 191, 198, 200, 201, 204
Secretion   9, 43, 192
Seeds   81
Segments   115
Segregation theory   145
SEIDEL, F.   113
Self-assembly   21–23
Seminal plasma   137
– vesicles   135
Seminiferous tubules   136
Sendai virus   195
Senescence   201
Serine   200
Serological methods   188
Sex chromosomes   135
– determination   66, 135
Sexual reproduction   155
SH/S-S ratio   59, 69, 74, 132, 137
Sheep   154
Siamese twins   31
Single-strand break   13
Sinistrality   86
Skeleton   121
Smittia   113
Sodium (Na$^+$)   72, 188
– pump   7
SO$_3$H groups   69
Somatic cells   54, 55, 59, 61, 95, 97, 98
Somites   121, 123, 125, 129, 130
Spacers   52, 54, 57, 163
SPEMANN, H.   3, 34
Sperm lysin   71
– nucleus   67, 69, 70, 95, 112, 166
Spermaster   67, 69, 70, 126
Spermatid   64
Spermatocyte   24
Spermatogenesis   25, 63, 135, 136
Spermatogonium   24, 136
Spermatozoon   19, 24, 25, 27, 44, 62, 64, 65, 67, 71–74, 79, 83, 84, 136, 137, 152, 172, 202, 205
Spermidine   103, 163
Spermine   103, 163
Spermiogenesis   64
S-phase   95
Spicules   116
SPIEGELMAN, S.   79
SPINDLE, A. I.   147
Spindle   57, 58, 61, 62, 82, 86, 91, 92, 94, 103, 169, 171, 177
Splicing   15, 57

S 100 protein   196
Stalk   155, 160, 163
Starfish   61, 62, 81
Stem cells   180, 183
Sterility   139, 172
Steroids   58, 59, 192
Sterols   124, 145
Strongylocentrotus   174
Subacrosomal space   64, 67
Sugars   175
Sulfate ions   117, 194
Suppressor   129
Symbiosis   18, 110
Synaptonemal complex   63

Tadpole   79, 83, 90, 100, 103, 110, 112, 128, 180, 182, 189
Tail   32, 90
TARKOWSKI, A. K.   141, 145
TATUM, E. L.   12
Taurine   137
Taxol   94
Telophase   91, 104, 105
Template   79, 97, 98
TENCER, R.   133
Teratocarcinoma   151, 180, 195, 202
Terminal differentiation   184
Testes   31, 135, 136, 149, 180
Testicular determinant (Td)   135
Testosterone   137
Tetracycline   18, 161
12-0-tetradecanoyl-phorbol-13-acetate (TPA)   198, 202
Tetraparental   145, 154
Tetraploidy   89, 171
Thalassemias   37, 194
THOMAS, C.   42
Threonine   200
Thymidine   75, 97, 98, 197
– kinase   162
Thymine   12, 13
TIEDEMANN, H.   4, 124, 130, 131
Tiger   153
Tight junctions   145, 148
Tissue affinities   185
t-mutants   178
Tobacco mosaic virus   21–23
TOIVONEN, S.   4, 124
Topoisomerases   20, 57
Totipotency   27, 146, 147
Toxic agents   36
Transcription   15, 18, 19, 38, 50, 57, 88, 101, 112, 129, 139, 140, 163–165, 169, 170, 178, 181, 197, 203

– factor   55
– units   49, 50, 52, 163
Transcriptional activity   152
– control   20, 37, 183, 192
Transcripts   115, 129, 142
Transdetermination   114, 191
Transdifferentiation   184, 191
Transfection   204
Transferrin   179
Transfilter induction   132, 192
Transforming factors   203
Transgenic mammals   178–180, 204
Translation   15, 18, 20, 56, 78, 80, 103, 113, 162, 164, 169, 170, 178
Translational control   37, 38, 109, 192
Transplantation   114
Transposable elements   115
"Trefoil" stage   28, 30, 168, 170
Tricarboxylic acid cycle   74
Tridermal animals   27
Triglycerides   194
Triplet   14
Triploidy   171
Trisomy   172
Triturus   49, 51, 53, 173
Trochophore   107
Trophectoderm   139, 142, 143, 145, 147, 148 ff., 152, 154, 178
Trophoblast   139, 142, 143, 148, 150, 153, 154, 178
Trypanosomes   14
Trypsin   69, 138, 192
Tubulin   7, 20, 65, 69, 78, 79, 89, 91, 94, 102, 103, 169, 177
Tumor   180, 191, 201–203
Tunicamycin   117, 133
Tunicate   110 ff.
Twins   205
Tyrosinase   111, 112, 196
Tyrosine   200

Ultracentrifugation   15, 54
Ultrastructure   3, 50, 51, 65, 68
Undermethylation (of DNA)   183, 201
Unfertilized eggs   57, 61, 69, 71–73, 76, 78, 81, 97, 99, 125, 141, 173, 176, 177, 191
Unipotent   143
Unique DNA sequences   13, 50
Uracil   12
Urea   94
Urethane   90
Uridine   88, 160
Urodeles   58, 69, 173
Uterus   59, 74, 81, 82, 135, 143, 154
UV microscope   6

UV-irradiation   31, 34, 113, 127, 166, 171
UV-sensitivity   127
Uvomorulin   145

Vacuole   158
Vagina   136
VAN ASSEL, S.   70
Vas deferens   136
Vectors   115
Vegetal pole   27, 32, 39, 41, 67, 87–89, 103,
    105, 110, 111, 115, 118, 119, 125, 126
Vegetalization   117
Vegetalizing substance, factor   117, 118,
    124, 125, 130, 131
Vegetative nucleus   158, 159, 162, 163, 165
Ventral lip (of the blastopore)   123, 128
Vertebrates   27, 34, 51, 115, 121, 128, 131,
    139
Villin   39
Vinblastin   94, 95
Viruses   10, 13, 21–23, 204
Visceral endoderm   151, 195
Vital dyes   131
Vitamins   4, 195
Vitelline membrane   39, 67
– wall   126
Vitellogenesis   39, 41–44, 47, 50, 52
Vitellogenin   41, 43, 192
Vitellolysis   175

WADDINGTON, C. H.   4
WARBURG, O.   3, 72
WASSARMAN, P.   141
WATSON, J. D.   13
WEHMEIER, E.   4

Wheat germs   204
WHITTAKER, J. R.   111
WICKENS, M. P.   57
Wolffian duct   136
– regeneration   189–191
Worm   81, 86, 106
WROBLEWSKA, J.   145
WURMSER, R.   3

X chromosome   135, 177, 205
– – inactivation   37, 152 ff., 201
Xenopus   34, 42, 45, 46, 51, 52, 54, 56,
    58–62, 72, 76, 80, 88, 100, 115, 128, 129,
    140, 165, 176, 196, 204
– borealis   44
– laevis   39, 40, 44
X-irradiation   89, 198

YAMADA, T.   4, 124, 190
Y chromosome   135, 205
Yeast   13
Yellow (A^y) mutant   177
Yolk   25, 27, 39, 83, 84, 86, 88, 106, 121,
    126, 128, 167, 175
– platelets   41, 43, 88, 100, 105, 109, 139,
    167, 175
– plug   29
– sac   151, 153

ZEUTHEN, E.   91, 92
Zona pellucida   138, 140, 147, 177
Zygote   66, 67, 75, 97, 135, 140, 141, 158,
    159, 162, 181, 182
Zygotene   24